Command and Control of U.S. Air Force Combat Support in a High-End Fight

DON SNYDER, KRISTIN F. LYNCH, COLBY PEYTON STEINER,
JOHN G. DREW, MYRON HURA, MIRIAM E. MARLIER,
THEO MILONOPOULOS

Prepared for the Department of the Air Force
Approved for public release; distribution unlimited

RAND PROJECT AIR FORCE

For more information on this publication, visit www.rand.org/t/RRA316-1

Library of Congress Cataloging-in-Publication Data is available for this publication.
ISBN: 978-1-9774-0665-1

Published by the RAND Corporation, Santa Monica, Calif.

Cover: U.S. Air Force photo/Tech. Sgt. Dana Rosso

Preface

The objective of this research was to recommend ways to make the command and control for U.S. Air Force combat support at the theater level more robust when operations are dispersed and communications are degraded. The work was conducted in fiscal year 2019 under a project entitled *Multi-Domain Command and Control for Agile Combat Support in a Degraded Environment* within the Resource Management Program of RAND Project AIR FORCE. The research was sponsored by the Director, Expeditionary Support, Air Force Installation and Mission Support Center, Air Force Materiel Command. It should be of interest to the combat support community and those who integrate command and control of logistics into the larger Department of the Air Force command and control processes.

RAND Project AIR FORCE

RAND Project AIR FORCE (PAF), a division of the RAND Corporation, is the Department of the Air Force's (DAF's) federally funded research and development center for studies and analyses, supporting both the United States Air Force and the United States Space Force. PAF provides the DAF with independent analyses of policy alternatives affecting the development, employment, combat readiness, and support of current and future air, space, and cyber forces. Research is conducted in four programs: Strategy and Doctrine; Force Modernization and Employment; Manpower, Personnel, and Training; and Resource Management. The research reported here was prepared under contract FA7014-16-D-1000.

Additional information about PAF is available on our website:
www.rand.org/paf/

This report documents work originally shared with the DAF on September 20, 2019. The draft report, issued on September 17, 2019, was reviewed by formal peer reviewers and DAF subject-matter experts.

Contents

Figures

Tables

Summary

This report proposes new concepts for the command and control of combat support to address the stresses presented by a high-end fight, specifically moving the location of the ground support for air operations to secure positional advantage against an adversary, which we refer to as *maneuver*,[1] and operations while under persistent multi-domain attack.[2] We do not assess whether maneuver is operationally effective or executable. We argue that these stressors present the following overall demands on combat support command and control:

1. decisionmaking and logistics direction that are timely and distributed when data, communications, or both are degraded
2. situational awareness that is timely and sufficiently comprehensive
3. coordination between combat support and operational activities that is timely and integrated
4. systems and processes for command and control that, when attacked across domains, are robust and resilient.

We compared the current joint and U.S. Air Force doctrine, policies, and recent deployment practices with these demands to reveal areas for improvements and recommendations to address these deficiencies. We found that

- for decisionmaking:
  - Combat support has no unified command and control mechanisms at the operational level.
  - The speed of combat support for maneuver is slowed down by the need for the coordinated actions of many actors, both within and outside the theater, including service, joint, and agency actors.
  - When issuing joint orders for the maneuver of air forces to new operating locations, the U.S. Air Force is hampered by its reliance on the air tasking order (ATO) for issuing orders.
  - The ability to adjust command and control of logistics when under persistent multi-domain attack is impeded by the lack of doctrine, policy, planning, and procedures for distributed command and control and for push logistics.[3]

---

[1] *Maneuver* is a generic concept that includes such concepts as adaptive basing, agile combat employment, and dynamic force employment.

[2] *Multi-domain attack* includes kinetic and nonkinetic attack. Both can degrade data and communications. Effects include loss of data, corruption of data, loss of access to data, and loss or degradation of the ability to communicate.

[3] A *push logistics system* does not wait for a demand from the user; it provides resources based on past consumption patterns and often includes buffer stocks to mitigate potential issues, such as incorrect forecasts or undersupply.

- for situational awareness:
  - The U.S. Air Force has not clearly defined the minimal information needed to maintain situational awareness for maneuver.
  - The U.S. Air Force lacks a common operating picture for combat support.
- for coordination between operations and combat support:
  - The separation of the bulk of combat support expertise and personnel (the U.S. Air Force forces [AFFOR] staff) from the locus of operational command and control (the Air Operations Center [AOC]) impedes coordination between operations and combat support.
  - The combat support teams in AOCs are currently understaffed.
  - The classification of some war plans presents a barrier to the logistics community for adequately planning for supporting operations.
- for robustness and resiliency of systems and processes:[4]
  - Functional fragmentation provides some robustness and resiliency to combat support operations.
  - The U.S. Air Force has limited deployable communications capabilities and capacity for robust and resilient communications.
  - The reliance on enterprise coordination makes operations fragile in a communications degraded environment.
  - The U.S. Air Force lacks policy for the prioritization of combat support information in a degraded data rate environment.

In sum, current combat support command and control processes and authorities are neither optimally designed for the timeliness needed for maneuver nor for the robustness and resiliency for operations while under persistent multi-domain attack.

Our principal recommendations to meet these needs are the following:

- Establish processes and support with appropriate organizational changes within the theater to issue a fragmentary order (FRAGORD) for logistics. We present options for placing the necessary organizational support for issuing a logistics FRAGORD under the Commander, AFFOR (COMAFFOR) and the Joint Forces Air Component Commander, such as
  - establish two directorates within the AOC:
    - Air Operations Directorate, which would replicate the AOC as it is currently constituted, with the five current divisions for generating thc master air attack plan and the ATO.

---

[4] By *robust*, we mean the protection of combat support–related information and the ability to operate combat support functions through the effects of dispersed operations and degraded communications. By *resilient*, we mean the ability to recover combat support functions satisfactorily during dispersed operations or degraded communications.

    - Airfield Operations Directorate, whose mission would be to create a logistics FRAGORD and direct maneuver.
  - establish, under the COMAFFOR, a commander under whom a logistics FRAGORD would be issued.
- Define doctrinal means by which command and control can devolve to the base level.
- Define minimal essential information for locations, both occupied and not occupied by the U.S. Air Force, for situational awareness to support decisionmaking for maneuver.
- Ensure that Airmen have access to the war planning information that they need to perform the logistics planning.
- Ensure multiple enterprisewide robust communications pathways for logistics and exercise their use.
- Define and follow procedures for backing up critical logistics data locally.
- Develop and exercise procedures for the command and control of push logistics.
- Establish the ability to provide a 24-hour, seven-days per week operational support cell within the Air Force Installation and Mission Support Center to support the warfighter during wartime.

These recommendations—which are not costly because they are changes to process (including organization) and doctrine, not material ones—would improve the resiliency and robustness of combat support command and control for a high-end fight.

Acknowledgments

We thank Brig Gen Brian Bruckbauer for his initiation of this work and his support throughout its execution. We also thank Marc VanDeveer and Col Kevin Heckle for their continued support.

We thank the numerous individuals in the U.S. Air Force who shared information and insights with us. They are too numerous to list individually. We thank the following organizations for their help: Air Combat Command, Air Education and Training Command Detachment 24, Air Force Installation and Mission Support Center, Air Force Petroleum Office, Air Force Security Forces Agency, Air Force Services Agency, Air Force Special Operations Command, Air Force Sustainment Center, Air Mobility Command, Pacific Air Forces, Headquarters U.S. Air Force (especially staff from the Deputy Chief of Staff for Logistics, Engineering, and Force Protection; the Checkmate staff; and Studies, Analysis, and Assessments), U.S. Air Forces in Europe, and the Defense Logistics Agency–Energy. Meetings with the U.S. Marine Corps were helpful for understanding how that service has addressed similar problems.

We especially thank Air Force Installation and Mission Support Center and U.S. Air Forces in Europe for allowing us to observe the Combat Support Wing and Rapid Forge exercises and for postexercise information and discussions.

At the RAND Corporation, we benefited from numerous discussions with Brien Alkire, Mahyar Amouzegar, Lt Gen (ret.) Robert (Bob) Elder, Jr., Jeff Hagen, Sherrill Lingel, and Patrick Mills. Reviews by Brien Alkire, Miranda Priebe, and William (Skip) Williams improved the report.

That we received help and insights from those acknowledged above should not be taken to imply that they concur with the views expressed in this report. We alone are responsible for the content, including any errors or oversights.

Abbreviations

AETF	Air Expeditionary Task Force
AETF/CC	Commander, Air Expeditionary Task Force
AEW	Air Expeditionary Wing
AFFOR	Air Force forces
AFIMSC	Air Force Installation and Mission Support Center
AOC	Air Operations Center
ATO	air tasking order
BOS-I	base operating support-integrator
COMAFFOR	Commander, Air Force forces
CONOP	concept of operations
CST	combat support team
DIRMOBFOR	Director of Mobility Forces
FRAGORD	fragmentary order
JFACC	Joint Force Air Component Commander
JFC	Joint Forces Commander
LTO	logistics tasking order
NATO	North Atlantic Treaty Organization
NEAF	Numbered Expeditionary Air Force
NEAF/CC	Commander, Numbered Expeditionary Air Force
NIPRNet	Non-Classified Internet Protocol Router Network
OODA	observe, orient, decide, and act
PACE	primary, alternate, contingency, and emergency
SIPRNet	Secure Internet Protocol Router Network
USEUCOM	U.S. European Command
USINDOPACOM	U.S. Indo-Pacific Command

1. Challenges to Combat Support

"If command and control fail, nothing else matters."

—Bruce G. Blair[5]

The Problem

Faced with a renewed emphasis on a near-peer fight, and after decades of post–Cold War military operations that barely threatened combat support,[6] the United States is refocusing its attention on investing in "forces that can deploy, survive, operate, maneuver, and regenerate in all domains while under attack." The implications for expeditionary combat support of this change in focus are to transition "from large, centralized, unhardened infrastructure to smaller, dispersed, resilient, adaptive basing . . ." and to "ensure logistics sustainment while under persistent multi-domain attack."[7]

This report proposes more-robust and more-resilient concepts for combat support command and control while under persistent multi-domain attack.[8] The main challenges that we address for combat support command and control are the abilities for the U.S. Air Force to maneuver[9] its ground posture in response to adversary threats and to maintain an acceptable level of combat support when communications capabilities are degraded or denied.[10] The discussion and recommendations in this report encompass an enterprisewide scope (not catered to specific geographic combatant commands) but are sensitive to the unique circumstances of each theater. Within the range of combat support, we concentrate on issues of fueling and arming aircraft, and

---

[5] Bruce G. Blair, *Strategic Command and Control: Refining the Nuclear Threat*, Washington, D.C.: Brookings Institution, 1985, p. 4.

[6] We use the term *combat support* in this report to mean what the U.S. Department of Defense calls *combat service support*: "The essential capabilities, functions, activities, and tasks necessary to sustain all elements of all operating forces in theater at all levels of warfare." *Logistics* is defined as follows: "Planning and executing the movement and support of forces" (see U.S. Department of Defense, *DOD Dictionary of Military and Associated Terms*, Washington, D.C., January 2020, pp. 40, 132). We will not distinguish between *combat support* and *logistics*.

[7] U.S. Department of Defense, *Summary of the 2018 National Defense Strategy of the United States of America: Sharpening the American Military's Competitive Edge*, Washington, D.C., 2018, pp. 6–7.

[8] By *robust*, we mean the protection of combat support–related information and the ability to operate combat support functions through the effects of dispersed operations and degraded communications. By *resilient*, we mean the ability to recover combat support functions satisfactorily during dispersed operations or degraded communications.

[9] By *maneuver*, we mean securing positional advantage against an adversary by moving the location of the ground support for air operations. Maneuver is a generic concept that includes such concepts as adaptive basing, agile combat employment, and dynamic force employment.

[10] By *degraded communications*, we mean the inability for the field to connect with critical information flows, reduced data rate between parties, or corrupted data.

maintenance and civil engineering support. Solutions focus on the near-term time horizon for supporting the current force structure, not potential next-generation weapon systems or novel concepts of operations (CONOPs). This near-term focus does not preclude prudently rethinking maintenance and civil engineering requirements for future combat systems and supporting equipment where appropriate.

Challenges for the command and control of combat support have been analyzed in past research.[11] Much of this work has focused on issues that arise during steady-state operations. We

---

[11] For past RAND research, see Robert S. Tripp, Lionel Galway, Paul Killingsworth, Eric Peltz, Timothy L. Ramey, and John G. Drew, *Supporting Expeditionary Aerospace Forces: An Integrated Strategic Agile Combat Support Planning Framework*, Santa Monica, Calif.: RAND Corporation, MR-1056-AF, 1999; Lionel Galway, Robert S. Tripp, Timothy L. Ramey, and John G. Drew, *Supporting Expeditionary Aerospace Forces: New Agile Combat Support Postures*, Santa Monica, Calif.: RAND Corporation, MR-1075-AF, 2000; Eric Peltz, Hyman L. Shulman, Robert S. Tripp, Timothy L. Ramey, and John G. Drew, *Supporting Expeditionary Aerospace Forces: An Analysis of F-15 Avionics Options*, Santa Monica, Calif.: RAND Corporation, MR-1174-AF, 2000; Robert S. Tripp, Lionel A. Galway, Timothy L. Ramey, Mahyar A. Amouzegar, and Eric Peltz, *Supporting Expeditionary Aerospace Forces: A Concept for Evolving the Agile Combat Support/Mobility System of the Future*, Santa Monica, Calif.: RAND Corporation, MR-1179-AF, 2000; Amatzia Feinberg, Hyman L. Shulman, Louis W. Miller, and Robert S. Tripp, *Supporting Expeditionary Aerospace Forces: Expanded Analysis of LANTIRN [Low Altitude Navigation and Targeting Infrared for Night] Options*, Santa Monica, Calif.: RAND Corporation, MR-1225-AF, 2001; Amatzia Feinberg, Eric Peltz, James A. Leftwich, Robert S. Tripp, Mahyar A. Amouzegar, Russell Grunch, John G. Drew, Tom LaTourrette, and C. Robert Roll, Jr., *Supporting Expeditionary Aerospace Forces: Lessons from the Air War over Serbia*, Santa Monica, Calif.: RAND Corporation, 2002, Not available to the general public; James Leftwich, Robert Tripp, Amanda Geller, Patrick Mills, Tom LaTourrette, C. Robert Roll, Jr., Cauley von Hoffman, and David Johansen, *Supporting Expeditionary Aerospace Forces: An Operational Architecture for Combat Support Execution Planning and Control*, Santa Monica, Calif.: RAND Corporation, MR-1536-AF, 2002; Don Snyder and Patrick Mills, *Supporting Air and Space Expeditionary Forces: A Methodology for Determining Air Force Deployment Requirements*, Santa Monica, Calif.: RAND Corporation, MG-176-AF, 2004; Mahyar A. Amouzegar, Robert S. Tripp, Ronald G. McGarvey, Edward W. Chan, and Charles Robert Roll, Jr., *Supporting Air and Space Expeditionary Forces: Analysis of Combat Support Basing Options*, Santa Monica, Calif.: RAND Corporation, MG-261-AF, 2004; Kristin F. Lynch, John G. Drew, Robert S. Tripp, and C. Robert Roll, Jr., *Supporting Air and Space Expeditionary Forces: Lessons from Operation Iraqi Freedom*, Santa Monica, Calif.: RAND Corporation, MG-193-AF, 2005; Patrick Mills, Ken Evers, Donna Kinlin, and Robert S. Tripp, *Supporting Air and Space Expeditionary Forces: Expanded Operational Architecture for Combat Support Execution Planning and Control*, Santa Monica, Calif.: RAND Corporation, MG-316-AF, 2006; Robert S. Tripp, Kristin F. Lynch, Charles Robert Roll, Jr., John G. Drew, and Patrick Mills, *A Framework for Enhancing Airlift Planning and Execution Capabilities Within the Joint Expeditionary Movement System*, Santa Monica, Calif.: RAND Corporation, MG-377-AF, 2006; Ronald G. McGarvey, James M. Masters, Louis Luangkesorn, Stephen Sheehy, John G. Drew, Robert Kerchner, Ben D. Van Roo, and Charles Robert Roll, Jr., *Supporting Air and Space Expeditionary Forces [AEFs]: Analysis of CONUS [Continental U.S.] Centralized Intermediate Repair Facilities*, Santa Monica, Calif.: RAND Corporation, MG-418-AF, 2008; Robert S. Tripp, William A. Williams, Kristin F. Lynch, John G. Drew, Dahlia S. Lichter, and Laura H. Baldwin, *A Strategic Analysis of Air and Space Operations Center Force Posture Options*, Santa Monica, Calif.: RAND Corporation, 2008, Not available to the general public; Kristin F. Lynch and William A. Williams, *Combat Support Execution Planning and Control: An Assessment of Initial Implementations in Air Force Exercises*, Santa Monica, Calif.: RAND Corporation, TR-356-AF, 2009; Ronald G. McGarvey, Robert S. Tripp, Rachel Rue, Thomas Lang, Jerry M. Sollinger, Whitney A. Conner, and Louis Luangkesorn, *Global Combat Support Basing: Robust Prepositioning Strategies for Air Force War Reserve Materiel [WRM]*, Santa Monica, Calif.: RAND Corporation, MG-902-AF, 2010; Kristin F. Lynch, John G. Drew, Amy L. Maletic, Robert S. Tripp, Ricardo Sanchez, William A. Williams, Brent Thomas, and Max Woodworth, *A Strategic Assessment of Component Numbered Air Force (C-NAF) Force Postures*, Santa Monica, Calif.: RAND Corporation, 2010, Not available to the general public; Robert S. Tripp, Kristin F. Lynch, John G. Drew, and Robert G. DeFeo, *Improving Air Force Command and Control Through Enhanced Agile Combat Support Planning, Execution, Monitoring, and Control Processes*, Santa Monica,

will not reiterate these issues in this report but focus instead on new issues that emerge or are exacerbated by the challenges posed when operations are dispersed or communications are degraded. We describe how dispersed operations and degraded communications affect combat support in the remaining sections of this chapter. In Chapter Two, we make key observations of the current command and control in this context to reveal new issues for combat support command and control, which are summarized at the end of that chapter.

Consequences for Combat Support Command and Control While Under Persistent Multi-Domain Attack

Maneuver

The ultimate goal of expeditionary combat support is to underpin and enable military operations in a theater. What needs to be supported, therefore, depends on specific CONOPs. Emerging CONOPs tend to incorporate some form of maneuver.[12] As we worked on this project in 2019, CONOPs that included maneuver as an element in survivable operations in a high-end fight were still evolving. Maneuver provides one way to respond to conventional threats by complicating adversary targeting—by moving capabilities from one location to another or by dispersing military capability across more locations. It also assists with survivability under limited nuclear threats by enabling forces to escape from radioactive plumes.

Given this evolution in concepts, and the need for combat support concepts to be robust to changes in CONOPs, we generalize the challenges that ground maneuver for the U.S. Air Force pose for combat support. As shown next, these general characteristics are not particularly sensitive to the details of CONOPs, which is to say that a wide variety of CONOPs involving maneuver will have the following characteristics. In this way, concepts for combat support command and control can be developed and evaluated as CONOPs evolve. If CONOPs are developed without the consideration of combat support consequences, it is possible that they would place demands on combat support so onerous as to be ineffective. In that case, CONOPs

---

Calif.: RAND Corporation, MG-1070-AF, 2012; Kristin F. Lynch, John G. Drew, Robert S. Tripp, Daniel M. Romano, Jin Woo Yi, and Amy L. Maletic, *An Operational Architecture for Improving Air Force Command and Control Through Enhanced Agile Combat Support Planning, Execution, Monitoring, and Control Processes*, Santa Monica, Calif.: RAND Corporation, RR-261-AF, 2014a; Kristin F. Lynch, John G. Drew, Robert S. Tripp, Daniel M. Romano, Jin Woo Yi, and Amy L. Maletic, *Implementation Actions for Improving Air Force Command and Control Through Enhanced Agile Combat Support Planning, Execution, Monitoring, and Control Processes*, Santa Monica, Calif.: RAND Corporation, RR-259-AF, 2014b; Patrick Mills, John G. Drew, John A. Ausink, Daniel M. Romano, and Rachel Costello, *Balancing Agile Combat Support Manpower to Better Meet the Future Security Environment*, Santa Monica, Calif.: RAND Corporation, RR-337-AF, 2014; and Kristin F. Lynch, John G. Drew, and Patrick Mills, *Enhancing Air Force Materiel Command Support to the Warfighter*, Santa Monica, Calif.: RAND Corporation, RR-2255-AF, 2018.

[12] CONOPs that include maneuver vary in frequency and timelines across a theater of operations.

are best adjusted so that combat support can accommodate them. Combat support and CONOPs for a high-end fight need to grow together, one influencing the other in an iterative process.

Forward Deployment

One concept of maneuver is the ability to rapidly deploy and employ forces at new expeditionary bases, even for short periods of time.[13] A scenario in which such a need might arise is when a main operating base is under imminent (or actual) kinetic attack. Aircraft might be moved to new locations to escape attack or to disperse the forces to complicate adversary targeting. Dispersed forces present an adversary with more targets to attack, and, if dispersion is conducted rapidly, force the adversary to locate and prosecute targets quickly. These moves reduce the danger of kinetic attack on aircraft on the ground (where they are most vulnerable) and impose additional costs on the adversary.

In this kind of maneuver, the newly occupied bases could very well be forward located, within kinetic attack range of the adversary. The strategy is to be close to the fight and protect the aircraft, not by withdrawing from the danger or defending against it but to survive by being mobile—staying inside the adversary's observe, orient, decide, and act (OODA) loop.[14] Commanders might seek alternative bases for refuge, or temporarily use forward operating bases for limited time intervals for refueling and rearming combat aircraft to increase the number of sorties and weapon deliveries. An exercise refueling and rearming F-15Es at an airstrip with minimal support and infrastructure was conducted in 2019 to test such a concept, called the Combat Support Wing concept.[15] The 388th Fighter Wing conducted another similar exercise in Europe, called Rapid Forge, in which aircraft were forward deployed to sites with minimal support.[16] Another similar scenario is to perform limited service on an aircraft away from its main operating base at a base closer to the fight to minimize transit time back to the home base.

In this set of cases, the salient characteristics for combat support are rapid movement in and out of a location, the ability to perform rapid service, and the ability to operate when under the

---

[13] For a recent review of Air Force maneuver, see Miranda Priebe, Alan J. Vick, Jacob L. Heim, and Meagan L. Smith, *Distributed Operations in a Contested Environment: Implications for USAF Force Presentation*, Santa Monica, Calif.: RAND Corporation, RR-2959-AF, 2019.

[14] The OODA loop concept was developed by John Boyd. The central idea is that the player, Red or Blue, that can observe, orient, decide, and act faster than the other gains the initiative and keeps the adversary confused. Boyd never formally published his work; for a fuller elaboration, see Lawrence Freedman, *Strategy: A History*, New York: Oxford University Press, 2013, pp. 196–201.

[15] A previous alpha test was conducted in July 2018. Brian Bruckbauer, Kevin Heckle, and Janelle Galang, *Combat Support Wing Capstone Exercise After-Action Report*, Joint Base San Antonio Lackland, Tex.: Air Force Installation and Mission Support Center, June 10, 2019, Not available to the general public; Michael Briggs, "Air Force Completes Test of Combat Support Innovation," Air Force News Service, May 14, 2019; and Stephen Losey, "Cops Refueling Jets: Small Teams, Sharing Jobs, Train to Set Up Airfields in Combat," *Air Force Times*, May 15, 2019.

[16] Kyle Cope, "Operation Rapid Forge Concludes," Air Force News Service, July 28, 2019; and Micah Garbarino, "Hill AFB Airmen Expand F-35A Combat Capability in Rapid Forge," Air Force News Service, July 19, 2019.

threat of kinetic attack (through all domains). If coordination of combat support and operations is not deliberate and well thought out, maneuver might not be effective in achieving its aims.

Retrograde Deployment

Maneuver might also be needed in retrograde deployment, again to seek refuge from kinetic attacks. In this case, the movement is away from the adversary, perhaps to take temporary refuge beyond the immediate threat reach of the adversary, at least until the kinetic threats to more forward main operating bases are sufficiently managed. The objective would be to do so without seriously degrading the combat effectiveness of the combat force. After the threats are sufficiently reduced, aircraft could then redeploy forward, closer to the fight. In this case, rapid movement is again a key characteristic (to keep the aircraft in the fight). An additional demand is the ability to adequately close the base that is being evacuated.

When moving retrograde, the U.S. Air Force needs to contend with cases of quickly departing from a large, established location with considerable equipment. A great deal of that equipment will likely need to be moved, not just to secure it from seizure, but because continuing operations might depend on it. Maintenance equipment, fuels trucks, and ammunition might all need to be moved. At the extreme end, one of those locations might include a central repair facility, whose capability would need to be reconstituted elsewhere. Closing bases, including maintaining security during drawdown, has recently proven to present significant challenges distinct from opening them.[17] One complicating factor in rapidly evacuating from a base will be the need to get aircraft out of a location, whether a forward location or a main operating base, that are too broken to fly. Repacking takes longer than unpacking and personnel are often more experienced and better trained in unpacking equipment than in packing it for movement.

Issues Specific to U.S. Indo-Pacific Command

Two features of the theater in U.S. Indo-Pacific Command (USINDOPACOM) present complicating factors to the maneuver of aircraft—great distances between runways and separation by water. The Pacific theater is large and the available main operating bases are often separated by great distances. The distance from Joint Base Pearl Harbor-Hickam (Oahu, Hawaii) to Kadena Air Base (Okinawa, Japan) exceeds 4,600 miles. The distance from Anderson Air Force Base (Guam) to Clark Air Force Base (Philippines) exceeds 1,600 miles, just a couple of hundred miles shy of the distance between Washington, D.C., and Salt Lake City, Utah. Distance translates to time in transportation, and these distances place high expectations on command and control to move forces quickly over these distances. Compounding this problem is that many of

---

[17] Although this referenced report focuses on the challenges of withdrawal from a theater and transition of security to a host nation, some of the issues, such as handing off local security and managing relations with the host country so that the location can be reused, are in common with the maneuver challenges. Michael M. Wellock, "Getting Out: Securing Air Bases During a Withdrawal," in Col Shannon W. Caudill, ed., *Defending Air Bases in an Age of Insurgency*, Vol. 2, Maxwell Air Force Base, Ala.: Air University Press, March 2019, pp. 87–109.

the airstrips in the Pacific theater are on islands, which often restrict movement of personnel and materiel by either air or sea. Although the fastest mode, movement by air is limited by available aircraft, maximum working capacity at airstrips, and air superiority. Shipping by sea is the least expensive mode, but it takes the longest time and can impose additional delays if intermodal transport is required. Even if the locations used in the maneuver tactic are beyond almost all intermediate-range adversary weapons, the maneuver location will almost certainly be attacked by nonkinetic means, at least indirectly. Nonkinetic means could have a high impact in the Pacific because of fewer alternative paths for connectivity among isolated islands.

Issues Specific to USEUCOM

The theater in U.S. European Command (USEUCOM) poses different complicating factors. Distances are shorter than in the Pacific and most sites can be accessed by landlines of communication. For command and control, the integration of U.S. forces with North Atlantic Treaty Organization (NATO) poses both advantages and disadvantages. The advantages are that numerous potential airstrips are available through partner nations and, because of the alliance, planning and diplomatic clearance can be conducted in advance. The concept of maneuver is not new to Europe and, although not a member nation of NATO, Sweden is reviving plans for dispersed basing and operating from austere locations.[18] The disadvantage in the European theater is the other side of the coin of the advantage—maneuver depends on the collective agreement of many NATO partners who will also be competing for the same airstrips and transportation.

Consequences of Maneuver

To summarize, whether moving forward or in retrograde, maneuver places several demands on combat support command and control. The principal demands are as follows:

- **The need to coordinate activities and needs between the operational and combat support communities.** Maneuver is used to complicate adversary targeting and facilitate combat operations. Combat operations and the threat environment will dictate where weapon systems are needed, the operational tempo that they need to meet, and the speed that is needed to stay inside the adversary's OODA loop. The physical limits of what resources are available and how rapidly they can move will constrain these operations. If combat support is not sufficiently coordinated with combat operations, plans might not be supportable and the mission could fail.
- **Rapid movement of personnel, equipment, and supplies, both in and out of operational locations.** Movement of forces must be fast enough to stay inside the adversary's OODA loop, including timely retreat to safety. For command and control, rapid movement includes the abilities to obtain diplomatic clearance in a timely manner, to make timely decisions about resource needs, and to execute logistical movements and

---

[18] Tony Osborne, "Sweden Is Relearning Cold War Basing Skills," *Aviation Week & Space Technology*, July 29–August 18, 2019, p. 20.

sustainment activities. Obtaining diplomatic clearance is not fully within the control of the U.S. Air Force. For Joint Task Force Noble Anvil[19] and during initial deployment for Operation Enduring Freedom,[20] diplomatic clearance could take weeks. In some cases, it was the longest step for achieving initial operating capability, exceeding the times to survey the site, build up the capability, and move in the forces.[21] An element of timeliness is prioritization—moving the most-critical resources for the operational mission first and relegating less-critical resources to later movement hastens the ability to meet operational mission needs.

- **Timely situational awareness of the conditions on the ground at (potential) operational locations.** Before moving, forces need to know that they are moving to a better situation and can perform the needed missions. At any candidate location, commanders will need situational awareness of the threat environment across domains, the nature of the site's infrastructure, the disposition of any personnel and equipment at the site, and an overall survey of the site. For combat support, this situational awareness is needed across all functional areas. In addition, information on other environmental factors will be needed, such as local weather and what resources might be available off the local economy.
- **Robust and resilient combat support operations while under the threat of kinetic attack and the ability to recover functions after attack.** In a high-end fight, combat support must be prepared to protect itself on the ground and be robust to the loss of infrastructure, equipment, and the wounding and killing of personnel. Upon attack, forces will need to be able to withdraw to safety or reconstitute capabilities. To recover capabilities after an attack, forces will need to allocate and distribute scarce resources. Personnel will need to backfill disabled personnel. Runway repair assets will be in demand. Civil engineers will be in demand to rebuild damaged infrastructure. Because resources may be needed in more than one location, decisions will be needed to prioritize allocation of resources and execute their movement and operations.

Although we will not focus on resourcing levels, if the U.S. Air Force maneuvers and is under attack, it will need supplemental resourcing levels of both personnel and equipment, both to fill the pipeline of movement and to be robust and resilient to combat loss.[22]

In a high-end fight, at minimum the U.S. Air Force will be part of a joint fight and all of these concepts and command and control will need to be integrated with the joint force.

Degraded Communications

Gone are the days of managing logistics by manual processing of orders and aircraft maintenance without computer processors. The information revolution has seen the automation

---

[19] Joint Task Force Noble Anvil was part of Operation Allied Force in Serbia in 1999.

[20] Operation Enduring Freedom was the initial operation in Afghanistan starting in late 2001.

[21] Robert S. Tripp, Kristin F. Lynch, John G. Drew, and Edward W. Chan, *Supporting Air and Space Expeditionary Forces: Lessons from Operation Enduring Freedom*, Santa Monica, Calif.: RAND Corporation, MR-1819-AF, 2004.

[22] Priebe et al., 2019, pp. 70–71.

of many combat support processes. Data are held, sometimes exclusively, digitally. Transactions are performed by computers using bespoke information technology. Diagnosis and management of aircraft maintenance depend on software intensive systems. Leaders maintain situational awareness and make decisions with the aid of digital decision-support tools. Manpower positions were cut as this automation occurred, and the remaining sleek workforce could not manually run the supply chain, diagnose the maintenance status and repair an aircraft, or conduct many other tasks without electronic equipment.

All of this information technology is vulnerable to cyber exfiltration and cyberattack. Beyond the cyber domain, the underlying communications infrastructure, such as undersea cables and space communications links, is also vulnerable to kinetic attacks, jamming, and other adversary impedance measures. It is not a mission of combat support to defend against kinetic and nonkinetic attacks. But it is the job of combat support to be able to carry out its functions in the face of such threats.

Cyber Exfiltration of Data

Although responsibility for defending networks against cyber exfiltration does not fall to the combat support community, it should shoulder some responsibility regarding its data. There is a great deal that the combat support community can do and not do to help secure its data and complement the actions taken by those who defend the networks. These actions are outside the topic of this report, and we will not elaborate on them.[23]

We emphasize that when combat support data are exfiltrated (or otherwise become known to an adversary), the adversary gains considerable situational awareness. That awareness could blunt the ability to survive by maneuver because the adversary has enough foreknowledge of the maneuver to get inside Blue's OODA loop. The success of rapid maneuver depends on staying inside the adversary's OODA loop. Therefore, the combat support community needs to pay attention to what can be gleaned, especially by sophisticated data aggregation and deep learning techniques, from the sum of its unsecured data. This adversarial opportunity will be a major factor in the "orient" part of the adversary's OODA loop. It will be a key factor in determining how quickly the U.S. Air Force must maneuver to be operationally successful and, therefore, how quickly combat support must be able to respond. It may also place current base and logistics information under higher operational sensitivities, creating a demand for operational-like data control strategies.

[23] Don Snyder, George E. Hart, Kristin F. Lynch, and John G. Drew, *Ensuring U.S. Air Force Operations During Cyber Attacks Against Combat Support Systems: Guidance for Where to Focus Mitigation Efforts*, Santa Monica, Calif.: RAND Corporation, RR-620-AF, 2015; and Don Snyder, Elizabeth Bodine-Baron, Mahyar A. Amouzegar, Kristin F. Lynch, Mary Lee, and John G. Drew, *Robust and Resilient Logistics Operations in a Degraded Information Environment*, Santa Monica, Calif.: RAND Corporation, RR-2015-AF, 2017.

Kinetic and Nonkinetic Attacks Against Data and Communications

Data and communications links are increasingly at risk of attack. Communications links can be severed. Data can be permanently erased. Data can be altered. Access to information systems and decision support tools can be denied. In the Pacific theater, islands are often serviced by a limited number of undersea cables with fewer landing points. These undersea cables supply most of their communications capacity. In early 2019, the failure of an undersea cable serving Tonga placed the island in a near-total internet blackout for more than a day.[24] The NotPetya cyberattack brought operations at the Danish shipping company Maersk to a near halt for several days and full recovery took weeks, and Maersk was not even the target.[25] The firm was able to recover because one computer in Ghana containing the company's domain controller data was unaffected by the attack (the computer was offline at the time of the attack in a coincidental, but ultimately fortunate, power outage). The information and communications technology infrastructures have real fragilities that put at risk the command and control of combat support, both at home and abroad.

Consequences of Degraded Communications

In the face of these threats, the combat support community needs concepts for continuing its operations when data are destroyed or altered, or access to data or the ability to communicate are delayed, curtailed, or denied. It is beyond our scope to assess the probability or extent of threats. Although being entirely cut off from data or communications links might be unlikely, losing specific data and access to dedicated information technology systems, such as supply systems, is a threat that must be addressed. Combat support must, at least temporarily, be able to operate through and recover from unreliable information, lack of information, and low data rates, including the inability to use specific data systems.

The principal demands on combat support are as follows:

- **The need for multiple, independent communications pathways and process-control systems to plan, execute, and manage combat support.** Single ways to communicate, control, or execute combat support operations are fragile to attacks in a high-end fight. A simple strategy for robustness and resiliency is to have independent means to accomplish each of these activities that are not susceptible to the same attack vector. That is, they do not share a common data processor, database, communication mode or path, and so on.
- **The ability, at least for brief periods, to distribute command and control of combat support.** When the ability to communicate is degraded, forces cannot rely on command and control by central authorities. To some extent, cutoff enclaves will need to carry out their missions as best as they can via conditional authorities and playbooks based on

---

[24] "Tonga Hit By Near-Total Internet Blackout," BBC News, January 23, 2019.

[25] Andy Greenberg, "The Code That Crashed the World: The Untold Story of NotPetya, the Most Devastating Cyberattack in History," *Wired*, September 2018, pp. 52–63.

previously issued commander's intent.[26] Deployed wings and groups will need to make local decisions regarding combat support within the theater. Reachback capabilities to enterprisewide support, such as the supply chain, will need to make and execute decisions regarding combat support to units for which they have limited or no communications.

- **The ability to operate temporarily with limited situational awareness.** Data and communications losses also mean that decisions will need to be made with less than a nominal understanding of the status of individual sites. Detailed knowledge of a deployed unit's spare parts needs, the status of its infrastructure, whether personnel have been wounded or killed, and other resource statuses might be meager or have delayed reporting. The U.S. Air Force will need to be prepared to operate under these less-than-ideal circumstances.
- **The ability to reallocate resources to recover nominal operations.** Recovery itself will require resources and the command and control of those resources to restore operations to nominal levels.
- **The ability to prioritize data and information flow.** All combat support is needed for the contingency operations, but not all combat support is equally critical in a given situation.[27] When communications are degraded, some prioritization will be needed to ensure that scarce communications capacity is allocated for the most-critical needs.

Summary of Goals

Several themes run through the challenges listed earlier. Maneuver while under "persistent multi-domain attack" places a high burden on command and control. The central themes indicate the following goals for combat support command and control:

- decisionmaking and logistics direction that are timely and distributed when data, communications, or both are degraded
- situational awareness that is timely and sufficiently comprehensive
- coordination between combat support and operational activities that is timely and integrated
- systems and processes for command and control that, when attacked across domains, are robust and resilient.

These demands pose core research questions for command and control of combat support that we address in this report:

- What should the decision process be for ground maneuver, and where should command authority lie?
- What should the process be for intra-theater movement of combat support?
- What organizational changes might be needed to facilitate effective maneuver?
- What is the minimal information needed for situational awareness of the ground to support maneuver?

---

[26] Priebe et al., 2019, Chapter 4; and Robert W. Peterman, *Mission-Type Orders: An Employment Concept for the Future*, Maxwell Air Force Base, Ala.: Air University, Air War College Research Report, March 1990.

[27] Snyder et al., 2015.

- Where would that information be stored, how would it be disseminated, and to whom?
- In times of degraded communications, how should information be prioritized, and by what process?

Premises and Bounds of Analysis

Underlying our analysis in this report are several premises and bounds of analysis. These premises and assumptions are points of departure for the analysis and will not be argued in depth.

Premises

Peacetime processes and organizations for the command and control of combat support should be designed to support wartime needs with minimal changes. One of the key lessons from World War II was that the price of reorganizing and developing new processes for command and control during wartime impeded effective operations.[28] When the exigencies of strategic airpower and intercontinental ballistic missiles compressed the timescale of a response to an attack to less than an hour, a single command and control system for both peacetime and wartime became necessary. Tactical warning, for example, went from being under national intelligence to being under operational control in the North American Aerospace Defense Command. To use the same command and control in both peacetime and wartime means, however, that some compromises will need to be made between peacetime efficiency and wartime effectiveness.

Shortcomings of command and control of combat support should not be the limiting factor in maneuver. This is to say that, if other factors, such as resource levels and time to move personnel and materiel, are sufficient for a desired level of maneuver, deficiencies in command and control should not be the restraining factor. Command and control should always be able to support what is otherwise operationally executable. The converse is that command and control does not need to support activities that are not feasible from an operational perspective.

Bounds of Analysis

In this report, although we use planned CONOPs to shape combat support command and control needs, we do not assess whether those CONOPS are operationally effective or executable. CONOPs for robust and resilient operations in a high-end fight were evolving at the time of the writing of this report. The desire for maneuver emerges largely from the current force structure, which drives the need to base and operate short-legged nonexpendable aircraft, such as fifth-generation fighters, near or within the kinetic threat range of an adversary. Other CONOPs include defensive measures for bases, denial and deception, and changing the force mix to long-

[28] David Jablonsky, *War by Land, Sea, and Air: Dwight Eisenhower and the Concept of Unified Command*, New Haven, Conn.: Yale University Press, 2010.

range aircraft or a larger number of expendable aircraft. Assessments will ultimately be needed of alternative CONOPs to determine which are operationally effective and whether those are executable. Concepts for combat support command and control will undergo a parallel evolution and will need to be assessed against each of the CONOPs. Ideally, the two will evolve iteratively. We examine the consequence of maneuver and do not evaluate the operational utility of maneuver.

We restrict analysis and recommendations to areas under the broad control of the logistics enterprise. The scope includes process and organizational changes at the enterprise level, within Air Operations Centers (AOCs); Commander, Air Force forces (COMAFFOR) A4 staffs; and maintenance and support groups (or air base wings). It is important to understand the scope of a command authority and whether it is administrative control through U.S. Air Force service equities or whether it requires a combatant command operational authority. We do not propose recommendations for hardening infrastructure or adopting new technologies or communications systems.

Analytic Approach

To address these research questions, we review the current concepts for the command and control of combat support in Chapter Two. That discussion is based on a thorough review of doctrine and policy at both the joint and U.S. Air Force levels. It is bolstered by examination of recent deployment practice by the U.S. Air Force and discussions with several individuals throughout the U.S. Air Force. Chapter Three compares the needs for combat support in a high-end fight in this chapter with the current state presented in Chapter Two to reveal specific findings of deficiencies. We then propose some potential courses of action to support maneuver, and more-general recommendations for more adaptive, robust, and resilient command and control for combat support.

2. Status Quo of Combat Support Command and Control

> "Organizations are hard to run; people don't always do what they are supposed to do. They also reflect diverse, conflicting external interests and diverse, conflicting internal interests. Information and knowledge is always insufficient, and the environment is often hostile and always somewhat unpredictable. Thus, there is the ever-present problem of prosaic, mundane organizational failure."
>
> — Charles Perrow[29]

In this chapter, we discuss how concepts for combat support are evolving in response to the National Defense Strategy, the status quo for combat support command and control, and then draw general observations about the limitations of current doctrine and policy.

Evolving Concepts in Combat Support

The combat support community has been exploring concepts for how it might support maneuver of air forces. Several exercises have tested aspects of these concepts, with the Combat Support Wing and Rapid Forge exercises being the most salient as of 2019.

Combat Support Wing Exercise

The Air Force Installation and Mission Support Center's (AFIMSC's) Combat Support Wing concept is a force deployment model to develop small, agile, multi-functional combat support teams (CSTs) for performing integrated combat turns with limited base support capabilities in austere environments.[30] Multi-functional Airmen are cross-trained to do multiple jobs on an aircraft (e.g., maintenance personnel who can help with security duties and security forces who can help with refueling).[31] In May 2019, AFISMC, in partnership with Air Combat Command, performed the capstone exercise for the Combat Support Wing concept. Teams of approximately 30 multi-functional Airmen deployed to three forward operating locations and performed integrated combat turns to refuel and rearm F-15E Strike Eagles while maintaining ground security. Refueling was done via trucks and no bomb building capabilities were deployed

---

[29] Charles Perrow, *The Next Catastrophe: Reducing Our Vulnerabilities to Natural, Industrial, and Terrorist Disasters*, Princeton, N.J.: Princeton University Press, 2007, p. 292.

[30] The description is largely based on the authors' observations of the exercise. See Bruckbauer, Heckle, and Galang, 2019; Briggs, 2019; and Losey, 2019.

[31] Three terms are being used in the U.S. Air Force: *Multi-functional Airmen* are ones who can work on a single mission-design series aircraft but can do multiple jobs; they are trained at the wing level. *Multi-qualified Airmen* are ones who can work multiple mission-design series aircraft but cannot sign off on work. *Multi-skilled Airmen* are ones who can work on multiple mission-design series aircraft, and are certified to sign off on the work. The latter would require a change in policy.

(uploaded bombs were already assembled and configured). Over the month prior to the exercise, Airmen received multi-functional training to execute security operations and perform other combat support tasks necessary for integrated combat turns.

Rapid Forge Exercise

In July 2019, U.S. Air Forces in Europe performed another distributed operations exercise named Rapid Forge.[32] During this exercise, F-15E Strike Eagles and F-35As were deployed from Germany to forward operating locations in Poland, Lithuania, and Estonia to perform integrated combat turns at austere bases. A single MC-130J Commando II aircraft was deployed to each forward operating location to provide the personnel, equipment, and supplies necessary to refuel, rearm, and perform limited maintenance on the fighter aircraft. Similar to the Combat Support Wing exercise, Airmen received multi-functional training to perform combat support tasks outside their Air Force specialty codes, and personnel and equipment were pre-positioned at each location in advance of the exercise. Command and control was also performed from a mobile facility in a simulated austere environment.

Rapid Forge marks the first exercise of F-35A aircraft for distributed operations and interoperability between fourth- and fifth-generation aircraft and with NATO allies. Rapid Forge has similarities to Rapid Raptor, which previously exercised the deployment of F-22 Raptors to austere locations and performed integrated combat turns with personnel, equipment, and supplies transported with a single C-17 aircraft.[33] Later in 2019, Agile Lightning also turned F-35As at forward locations in an exercise similar to Rapid Forge.[34]

Implications of Evolving Concepts

These concepts and exercises are designed to demonstrate the feasibility of rearming and refueling fighter aircraft at austere locations. During wartime and under multi-domain attack, these concepts place new demands on combat support command and control. A notional example illustrates some of the demands.

Suppose a small package of fighter aircraft were to rearm and refuel for a limited time (e.g., hours, a day, or perhaps several days) from a forward operating location. Advanced planning and preparation would be needed, including securing permission from the host nation to operate out of the location. Planners would need a site survey sufficiently detailed, accurate, and up to date. This advanced planning would be followed by further planning and execution of the movement

---

[32] The description is largely based on the authors' observations of the exercise. See also Garbarino, 2019; and Cope, 2019.

[33] Marc V. Schanz, "Rapid Raptors: A New PACAF Concept Gets F-22s to the Fight Fast," *Air Force Magazine*, November 2013, p. 57.

[34] Staci Kasischke, "Agile Lightning Demonstrates Nimble Operations," U.S. Air Forces Central Command Public Affairs, August 12, 2019.

of combat support resources to the site, including personnel, equipment, and supplies. Full-up rounds must be shipped to the site for rearming, or, for bombs needing assembly and configuration, equipment for bomb building must be moved to the site (otherwise, full-up rounds of bombs must be assembled elsewhere and moved in a fully assembled state). For refueling, access to a fuel source is secured locally (typically negotiated by the Defense Logistics Agency), stored at the site, and delivered to the aircraft on demand (e.g., via a refueling truck). For the duration that resources, such as a truck or storage bladder, are in transit, they are unavailable, filling a supply pipeline. The more movement, the larger the pipeline and the larger the force needed in theater to meet operational needs.

Many questions arise. Who will make the commands to vacate a forward operating location and maneuver elsewhere? Who will have the authority to reallocate resources from one wing to another? How will these orders be issued? Contingency plans will be needed for what to do when an aircraft breaks on the ground (or for some reason engines must be shut down and no aerospace ground equipment is on site to restart the engines). How will maintenance issues be diagnosed and repaired under these conditions? How will the site be secured if aircraft are disabled on the ground for protracted periods? Under what circumstances will the aircraft be destroyed if physical security is weak? Who will give this command? Who will prioritize these support activities relative to other demands in the theater? How will these priorities be issued as commands?

Expeditionary Command and Control: Status Quo

Within a geographic combatant command, the combatant commander is responsible for all missions assigned to the command.[35] Although doctrine at the joint and U.S. Air Force levels specify canonical organizational structures and command relationships, the combatant commander has discretion to organize his or her forces and often exercises this right to cater command structures to evolving circumstances.[36] Doctrine and practice define two typical overlapping roles within the command structure.[37]

The first is that the forces provided by a service are generally organized under a service component command. In the case of the U.S. Air Force, forces assigned to the geographic combatant commander are under the unified command of the COMAFFOR who is responsible to the geographic combatant commander reporting to the Secretary of Defense. The COMAFFOR has the responsibility to represent the service and employ U.S. Air Force forces to accomplish

---

[35] U.S. Code, Title 10, Section 164, Commanders of Combatant Commands: Assignment; Powers and Duties.

[36] Doctrine specifies typical or idealized organizational structures and is not prescriptive. Each geographic combatant commander organizes somewhat differently. Therefore, actual command structures generally differ in detail with doctrine. We present doctrinal structures in this section because they serve as a baseline for discussion.

[37] See, for example, Joint Publication 3-30, *Joint Air Operations*, Washington, D.C.: Joint Chiefs of Staff, July 25, 2019; and Air Force Doctrine Annex 3-30, *Command and Control*, Maxwell Air Force Base, Ala.: Curtis E. LeMay Center for Doctrine and Education, November 7, 2014.

assigned functions and tasks, duties which include service logistics support for all U.S. Air Force forces within his or her geographic region. The COMAFFOR might not have strict administrative control for everyone in the theater, especially if they represent a global enterprise force capability.

The second is a functional division of command along the lines of land, maritime, special operations, and air. Joint air operations are under the unified command of the geographic combatant commander and, if named, a Joint Force Air Component Commander (JFACC).[38] The JFACC "plans, coordinates, allocates, tasks, executes, and assesses joint air operations to accomplish assigned operational missions."[39] Both the COMAFFOR and the JFACC either report directly to the combatant commander, or, if there is more than one operation in the theater, to a Joint Forces Commander (JFC) who reports directly to the combatant commander, as depicted in Figure 2.1.[40] However, generally the COMAFFOR is appointed through service channels and the JFACC designated by the JFC.

Figure 2.1. Doctrinal High-Level Command Structure

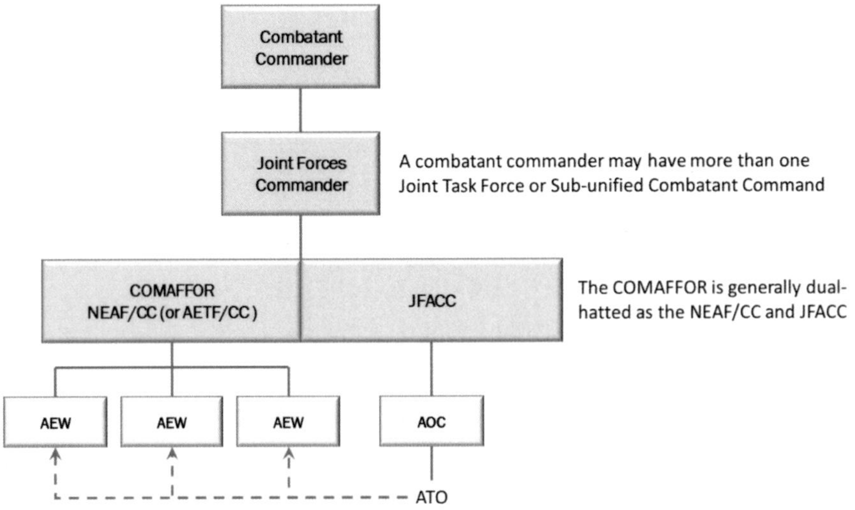

SOURCE: Air Force Doctrine Annex 3-30, 2014; Joint Publication 1, *Doctrine for the Armed Forces of the United States*, Joint Chiefs of Staff, incorporating change 1, July 12, 2017.
NOTES: Solid lines represent operational command reporting and dashed lines represent orders. A Numbered Expeditionary Air Force (NEAF) is an alternative name for an Air Expeditionary Task Force (AETF) when the AETF is made up of multiple wings. AETF/CC = Commander, Air Expeditionary Task Force; AEW = Air Expeditionary Wing; ATO = air tasking order; NEAF/CC = Commander, Numbered Expeditionary Air Force.

---

[38] When operations include allied countries, allied airpower falls under the unified command structure and the JFACC is called a Combined Forces Air Component Commander. We will use the term *JFACC* throughout this report, keeping in mind that operations could be combined. Some air operations do not typically fall under the command of the JFACC (e.g., rotary-wing air support to ground troops and aircraft dedicated to the defense of naval ships).

[39] Air Force Doctrine Annex 3-30, 2014, p. 7.

[40] A single JFACC or COMAFFOR can support multiple JFCs.

The COMAFFOR

Doctrine specifies that the COMAFFOR commands at the operational level of war and is the unified commander of all U.S. Air Force forces reporting to a Joint Forces Commander. The COMAFFOR typically maintains operational control over assigned Air Force forces. The COMAFFOR is always a U.S. Air Force general; a member of the service with the preponderance of air forces is normally assigned the role of JFACC.[41] There have been recent cases in which the COMAFFOR and the JFACC were not the same individual and the JFACC was not American.[42] If these commands are separate, integrated command of air forces is more complicated. In a high-end fight, we expect that a large part of the U.S. Air Force will be engaged, so that the preponderance of air forces will be from the U.S. Air Force and the COMAFFOR will be dual-hatted as the JFACC. In this report, we assume that a single commander is dual-hatted in these two roles; however, we also point to some implications if he or she is not dual-hatted and present options to deal with this case.[43]

All U.S. Air Force forces report to the COMAFFOR, who is supported by an AFFOR staff. Within that AFFOR staff, combat support responsibilities are spread across the A1 (mortuary affairs, services), A4 (maintenance, supply, fuel), A6 (communications), and A7 (mission support, force protection, civil engineering, emergency response).[44] The role of the AFFOR staff in the context of maneuver is to coordinate combat support within the theater for the COMAFFOR.[45] The staff have no direct operational or tactical command authority over combat support.

Typically, the U.S. Air Force forces under a geographic combatant commander are organized as an Air Expeditionary Task Force (AETF), or, in large operations with multiple wings, a Numbered Expeditionary Air Force (NEAF).[46] The COMAFFOR serves as the commander of the AETF or the NEAF. For a high-end fight, it is possible that the number of forces will be large enough to be organized into more than one Air Expeditionary Wing (AEW) and therefore a NEAF rather than AETF structure is expected, but none of the conclusions that we reach rest on this assumption.

---

[41] Joint Publication 3-30, 2019.

[42] Examples include operations in Kosovo in the late 1990s (see Lynch et al., 2005) and Libya in 2011 (see Karl P. Mueller, ed., *Precision and Purpose: Airpower in the Libyan Civil War*, Santa Monica, Calif.: RAND Corporation, RR-676-AF, 2015).

[43] In 2019, both U.S. Air Forces in Europe and Pacific Air Forces reorganized to place the AOC under the AFFOR/A3, which complicates matters if the same individual is not dual-hatted as the COMAFFOR and JFACC.

[44] A-staffs can be combined (e.g., A4/7), but in no case has there been an A1/4/6/7 that would combine all combat support.

[45] Air Force Instruction 13-103, *AFFOR Staff Operations, Readiness, and Structures*, Washington, D.C.: Secretary of the Air Force, August 19, 2014.

[46] The organizational structure of a NEAF is defined in Air Force Doctrine Annex 3-30, 2014, p. 58.

The JFACC

The JFACC operates at both the operational and tactical levels of war and is typically delegated tactical control of the tasked forces from the Joint Forces Commander.[47] Although the JFACC can have a small staff, the JFACC exercises command and control of air operations through an AOC.[48] The AOC was developed to control air operations at the operational level of war. It typically has divisions for strategy; combat plans; combat operations; intelligence, surveillance, and reconnaissance; and air mobility. It does not have a division for combat support.

One key product of the AOC is the air tasking order (ATO).[49] The ATO is a detailed product that specifies the number of sorties, refueling tracks and times, targets, times over target, selection of ordnance, coordinating and controlling activities, and assigning communications frequencies. The ATO serves a similar role for air operations that fragmentary orders (FRAGORDs) serve for other joint operations.

In the joint planning process, operation orders are used as directives to subordinate commanders to affect the coordinated execution of an operation.[50] FRAGORDs are then used to modify previously issued orders and provide either brief instructions on the parts that have changed or greater detail for the execution of the operation. However, the U.S. Air Force does not generally use this aspect of the joint construct for generating orders. U.S. Air Force policy and doctrine rarely mention FRAGORDs. When FRAGORDs are mentioned, it is either in the context of joint planning or simply to acknowledge their existence within the joint construct.

The ATO is issued daily, with a three-day planning cycle. Some limited activities must be planned and executed faster than the ATO cycle, one example being dynamic targeting (time-sensitive targeting). These faster activities are handled by special cells in the AOC and coordinated with the ATO.

Combat support, on the other hand, has not developed an operational-level warfighting construct with roles and responsibilities similar to the AOC. Generally, when the JFACC and COMAFFOR are the same individual, combat support issues may be worked out within the COMAFFOR staff (or by a numbered air force or major command staff associated with the geographic area). Therefore, the challenge is live collaboration with the AOC planning and ATO production during a conflict. Over the years, there have been ad hoc attempts to fill some of this role. Air Force Materiel Command established a warfighting cell during Operation Enduring Freedom (Afghanistan), for example, but there is no standing organization for combat support

[47] Joint Publication 3-30, 2019.

[48] When overseeing joint operations, the AOC is called a Joint Air Operations Center, and when overseeing combined operations, a Combined Air Operations Center. For simplicity, we conform to doctrine in using the generic term *AOC*, recognizing that it may be directing joint or combined forces.

[49] The AOC also produces the master air attack plan and the area air defense plan.

[50] Joint Publication 5-0, *Joint Planning*, Washington, D.C.: Joint Chiefs of Staff, June 16, 2017.

equivalent to the AOC. Within the AOC, a provision is made for a CST to be the focal point for combat support in the AOC. If so empowered, this team, when constituted, could direct maneuver on behalf of the JFACC (or JFC).[51] It is, however, a small cell, not a division, and in general needs outside assistance from the AFFOR staff to understand the combat support implications of beddown decisions, much less orchestrate those movements.[52] Current AOCs in USINDOPACOM and USEUCOM, however, only staff CSTs when they determine they are needed.

Maneuver

Maneuver is a decision executed at the tactical level of war but bears operational and strategic implications. The tactical level involves details of the immediate threat environment, weather conditions, and other factors that influence decisions on whether and at what time to either accept or avoid risk at one location or another. But movement of ground support or basing of aircraft cannot be conducted in isolation. To support maneuvering forces, combat support resources need to move from one location to another. During movement, those resources reside unused in a supply pipeline, with implications for capabilities elsewhere in the theater—and, therefore, with effects at the operational level of war. Furthermore, access to new locations and permissions for changes to operations out of existing bases require host-nation permission and must be done with the assistance of diplomatic outreach. This aspect of maneuver is inherently strategic.

When a JFC is established, the ultimate operational command authority for an operation within the theater, including that for the maneuver of forces, lies with the JFC.[53] The JFC could delegate decisions for the maneuver of U.S. Air Force forces to a subordinate commander, such as a JFACC, who could be dual-hatted as a COMAFFOR and further delegate some decisions to lower echelons as circumstances indicate. Whether a commander is delegated these authorities in their role as JFACC or COMAFFOR is less important than the capability that that commander possesses regarding the processes, procedures, and coordination mechanisms to support maneuver decisions. If the COMAFFOR and JFACC are not the same individual, it is even more critical that organizations under the control of the delegated commander are sufficient to the task. In the event that the JFC delegates the authority to maneuver to lower echelons, we have not found guidance for such delegation of the authority to move the beddown or transient operating location of aircraft in doctrine or policy.

---

[51] A change in Air Force Instruction 13-1AOC would be required to empower the CST to act as more than a liaison cell between the AOC and the AFFOR staff (see Air Force Instruction 13-1AOC, *Operational Procedures: Air Operations Center (AOC)*, Washington, D.C.: Secretary of the Air Force, November 2, 2011, incorporating change 1, May 18, 2012).

[52] For more information, see Section 8.10 in Air Force Instruction 13-1AOC, 2012.

[53] Joint Publication 3-0, *Joint Operations*, Washington, D.C.: Joint Chiefs of Staff, incorporating change 1, October 22, 2018.

Combat Support to the COMAFFOR

Combat support is divided into 25 functional communities.[54] Although some combat support functional communities were consolidated through the establishment of Air Force Sustainment Center and AFIMSC, management of combat support capabilities is still divided among and between organizations both inside and outside the U.S. Air Force (see Figure 2.2).

Figure 2.2. Major Combat Support Coordination Entities for AFFOR Staff

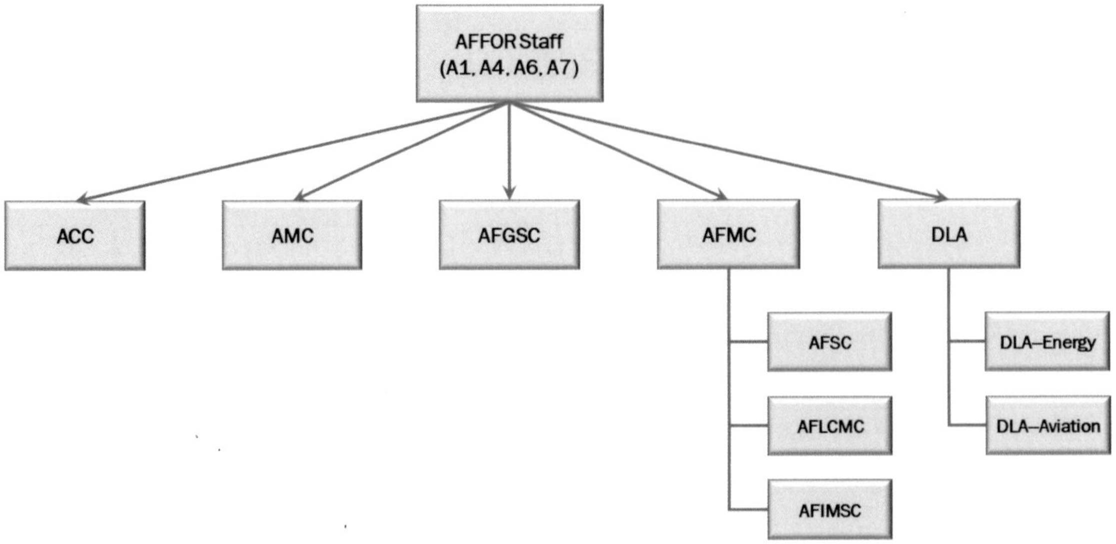

NOTE: AFSC = Air Force Sustainment Center.

Combat support capabilities are provided in several different ways, through unit-owned assets or capabilities, war reserve materiel, host-nation support, or contracted assets or capabilities. As an example, the AFIMSC manages unit-owned civil engineering assets, whereas civil engineering war reserve materiel responsibilities are divided between the Air Force Sustainment Center (which is responsible for most of the civil engineering war reserve materiel) and the AFIMSC (which is responsible for various items, such as rations). And, the AFFOR staff is responsible for coordinating host-nation civil engineering support.[55] Managing combat support capabilities in steady state with assured communications is complex, involving many

---

[54] These functional communities are Acquisition; Air Force Office of Special Investigations; Airfield Operations; Analyses, Assessments, and Lessons Learned; Chaplain Corps; Civil Engineer; Communications/Information; Contracting; Distribution; Force Support; Financial Management/Comptroller; Health Services; Historian; Judge Advocate; Logistics Planning; Maintenance; Materiel Management; Munitions Management; Postal Services; Public Affairs; Safety; Science/Technology; Security Forces; Test and Evaluation; and Weather Services (see Air Force Doctrine Annex 4-0, *Combat Support*, Maxwell Air Force Base, Ala.: Curtis E. LeMay Center for Doctrine and Education, December 21, 2015).

[55] See Lynch, Drew, and Mills, 2018.

organizations. Maneuver, whether moving between existing bases or dispersing to new bases, only further complicates the management of these capabilities.

The combat support functional capabilities most heavily relied on for maneuver include airfield operations, civil engineer, communications, contracting, distribution, health services, logistics planning, maintenance, materiel management, munitions management, security forces, and weather services. For the command and control of resources, we focus on three categories: the personnel, the equipment, and the supplies.[56] The assignment and movement of these three categories to a beddown location in a theater are generally handled differently. Then, where those resources are located (from where they are sourced) also affects how they are handled. There are different processes if the resources are moved within the theater, from resources assigned or allocated to another theater, or from resources neither assigned nor allocated to another theater.

Resource Management

Many processes combine to get the right resources to each unit that needs them at the right place and time. The point of this section is to describe several of these processes to highlight the number of players involved and degree of coordination that is typical of managing logistics in a theater. These processes tacitly evolved over time in part for effectiveness, in part for efficiencies, and in part for accountability of resources beyond the theater level. These processes are not specifically tailored to be agile with respect to maneuver.

Consider first the reallocation of resources within a theater. Directive authority for logistics grants authorities for reallocating any resources within a theater to the combatant commander.[57] For resources that lie wholly within the U.S. Air Force, the combatant commander delegates authorities to the COMAFFOR. For aircraft squadrons, this is commonly practiced using an ATO issued by the AOC. There is no similar common practice or single order to reallocate combat support resources within the theater.

When an AEW needs resources, combat support planners submit their requirements to the AFFOR staff for coordination and sourcing. If the requirements can be sourced within the theater, the AFFOR staff can request that wings redistribute their resources to fulfill the requirements. However, the AFFOR staff is a staff, not an execution organization. It does not have the authority to direct and control resource allocations, so disputes would need to be

---

[56] For the purposes of this report, we refer to Supply Class 1 (subsistence, including food), Supply Class 3 (petroleum, oil, and lubricants), Supply Class 5 (ammunition), Supply Class 7 (major end items), Supply Class 8 (medical materiel), and Supply Class 9 (repair parts).

[57] 10 U.S.C. § 164; and Joint Publication 4-0, *Joint Logistics*, Washington, D.C.: Joint Chiefs of Staff, incorporating change 1, May 8, 2019.

resolved by the COMAFFOR, who, as noted previously, does not have processes or procedures to exercise tactical-level command.[58]

If the requirements are to be sourced from another service component within the theater, the AFFOR staff submits a request to the geographic combatant command to redistribute the resources. Logistics at the joint level is coordinated by the combatant command through a Joint Logistics Operations Center under the combatant command J4. The AFFOR staff coordinates most logistics activities with the Joint Logistics Operations Center.[59]

Supplies, equipment, and personnel are, in general, managed by separate processes and control systems. The management of classes of supply is fragmented by functional areas, each with independent databases and process-control systems. Some of these databases and process-control systems are U.S. Air Force enterprise systems, and some extend beyond the U.S. Air Force (e.g., Defense Logistics Agency). At the enterprise level, each supply class is managed by one or more global resource managers or global force managers. Fuel, for example, is provided for each branch of the military across the globe by the Defense Logistics Agency–Energy. Individual management of combat support functional areas allows processes and systems to be tailored to specific needs and may promote greater efficiency. But it means that movement of supplies, even within the theater, cannot be done independently of the enterprise.

Equipment and personnel, although managed separately from supplies, share the common attribute of generally needing some reachback to enterprise systems for accountability and process execution.

Movement of resources within a theater is generally managed under the combatant commander by a Joint Deployment and Distribution Operations Center (JDDOC).[60] A Joint Transportation Board adjudicates conflicts in priorities. The JDDOC develops deployment and distribution plans and coordinates distribution within the theater above the tactical level. Tactical airlift assigned to the theater is allocated within the AOC by the Director of Mobility Forces (DIRMOBFOR). The DIRMOBFOR exercises coordinating authority for intra-theater and inter-theater movement of resources. The Army's Theater Sustainment Commands in USINDOPACOM and USEUCOM are also responsible for inland transportation (e.g., truck, rail, barge) at the tactical level of war using both U.S. Department of Defense and commercial transportation assets.[61]

---

[58] Unlike the JFACC, who has tactical control and thus the authority to use the AOC to direct the assigned wings to accomplish the mission, the COMAFFOR and AFFOR staff do not doctrinally exercise tactical control or directive authority of combat support resources (see Joint Publication 1, 2017).

[59] Joint Publication 4-0, 2019.

[60] Joint Publication 4-1, *The Defense Transportation System*, Washington, D.C.: Joint Chiefs of Staff, July 18, 2017.

[61] The U.S. Army's 8th and 21st Theater Sustainment Commands support USINDOPACOM and USEUCOM, respectively.

Now consider the allocation of resources to a theater from outside the theater. Although maneuver might be faster if most of the necessary resources for support were in theater, some of the resources might have to come from outside the theater. If coming from outside the theater, requests for equipment and personnel go through the same processes and actors as those discussed earlier, specifically the AFFOR staff and Joint Logistics Operations Center. (Some supplies can be requested by a unit directly to the enterprise system, but others, such as rations, will flow through the AFFOR/A1.) The combatant commander then forwards the request for forces to the Office of the Secretary of Defense. It then flows to the U.S. Air Force for sourcing. If resources are requested from another combatant command, the request might need to be resolved at the Chairman of the Joint Chiefs of Staff or Secretary of Defense level.

Movement, and the prioritization of this movement, is then managed through a time-phased force deployment data system controlled by, in the Department of the Air Force, the Deliberate and Crisis Action Planning and Execution Segments system and, at the joint level, by the Joint Operation Planning and Execution System. Movement is executed by U.S. Transportation Command. Securing transportation can be a lengthy process that is subject to formal review and validation of requirements using a well-defined prioritization framework.[62]

Whether resources are reallocated within the theater or arrive from outside the theater, a quick flow of resources to a location to support maneuver places high demands on the synchronization of prioritization. To make the combat support work, each activity (e.g., aircraft movement, infrastructure repair after attack) must be prioritized in a coordinated manner, each resource (across supplies, equipment, and personnel) must have an appropriate prioritization to support the activities, and the movement must be appropriately prioritized so that resources arrive at the proper juncture. The activities, resource allocation, and movement also all need to be mutually prioritized and synchronized. Any of these factors falling out of synchronization with the others can lead to poorly used resources or mission failure. No command and control exists to mutually prioritize and synchronize these processes; existing authorities are principally spread across the COMAFFOR, AFFOR staff, JDDOC, DIRMOBFOR, and Joint Transportation Board.

Installation Management

A base typically has an installation commander who is responsible for the protection of forces and assets on the base and for the lodging, dining, and administrative reporting of the personnel. The installation commander is also responsible for local host-nation support. Doctrine states a preference that the installation commander be an operational commander of a wing, group, or (no lower echelon than a) squadron.[63] The installation commander can be the same as

---

[62] Chairman of the Joint Chiefs of Staff Instruction 4120.02D, *List of Priorities—DoD Transportation Movement Priority System*, Washington, D.C., May 3, 2019.

[63] Air Force Doctrine Annex 3-30, 2014.

the combatant commander–designated officer responsible for providing base operating support, the base operating support-integrator (BOS-I).[64] The BOS-I is responsible for base facilities layout and construction, often coordinating capabilities provided by different services used to construct the operating location.

Airfield sites also have an officer in charge of the "control, operation, and maintenance of the airfield to include the runways, associated taxiways, and parking ramps as well as land and facilities affecting airfield operations," called the *senior airfield authority*.[65] BOS-I responsibility is typically assigned to the service with the preponderance of forces at an installation. The BOS-I might be drawn from the Army and the senior airfield authority from the U.S. Air Force. Joint policy anticipates that, in many cases, the contingency base commander will be the same as the BOS-I.[66]

To plan for operations out of a new site, planners need information about the infrastructure, runways, topography, and other details to know what additional resources are needed, how to secure the site, and how to posture forces on the site. This preparation takes the form of base support plans. By policy, plans are developed only for sites that explicitly appear in approved joint war plans. Joint war plans can carry classification levels that limit the number of logistics personnel who can access them, hampering this advanced planning.

Communications

It is expected that adversaries will target communications by both kinetic and nonkinetic means. Responsibility for tactical communications at deployed locations falls to combat communications and contingency response units; they are responsible for communications for 45 days when opening a base. After that, combat communications units are responsible for providing communications for up to 180 days or until the forward operating location can be transitioned onto commercial infrastructure and expeditionary communications forces arrive. At that juncture, combat communications units are then available to be redeployed to another operating location.

Communications flyaway kits are the primary method by which combat communications and contingency response units provide tactical communications to forward operating locations. Communications flyaway kits are compact, portable communications packages designed for rapid deployment and take 30 minutes to set up.[67] Communications flyaway kits provide access to Non-Classified Internet Protocol Router Network (NIPRNet), Secure Internet Protocol Router

---

[64] Air Force Doctrine Annex 4-0, 2015; and Joint Publication 4-04, *Contingency Basing*, Washington, D.C.: Joint Chiefs of Staff, January 4, 2019.

[65] Air Force Doctrine Annex 4-0, 2015, pp. 21–22.

[66] Joint Publication 4-04, 2019.

[67] Communications flyaway kits contain three ruggedized cases that weigh 30 pounds to 45 pounds each.

Network (SIPRNet), and Voice over Internet Protocol using commercial and military communication satellites, cellular networks, Wi-Fi, and wired communications infrastructure.[68]

Communications flyaway kits are also often components of primary, alternate, contingency, and emergency (PACE) communications plans for forward operating locations (see Table 2.1 for an example). PACE plans are meant to document how units plan to communicate when operating in a contested environment. However, PACE plans are fairly nascent and still developing. One weakness of current PACE plans includes a significant dependence on commercial infrastructure and services at all levels for both forward operating locations and main operating bases, placing the defense in depth at risk.

Reliance on commercial systems exposes military combat support to additional risks over military systems. Commercial logistics systems do not need to comply with military security standards. There is little market incentive to secure commercial systems to the highest level of nation-state attacks, whether kinetic or nonkinetic. More importantly, large commercial logistics firms are international, opening them to potential insider access by U.S. adversaries. A potential adversary might supply information technology systems to commercial firms that can be compromised at will. And global logistics firms do not screen employees by loyalty to U.S. interests and therefore might employ personnel with sympathies to a potential adversary, posing a direct insider threat. As a result, the Department of the Air Force has no control over and little visibility into the security of commercial firms, expanding its risk exposure.

Table 2.1. Notional Example PACE Plan

PACE Level	Communication Method
Primary	Terrestrial fiber communications
Alternate	Large aperture satellite communications
Contingency	Communications flyaway kit
Emergency	Secure Iridium satellite phone

SOURCE: 5th Combat Communications Group, *5 CCG Planners Guide*, May 30, 2018.

As mentioned earlier, logistics processes often rely on enterprise systems, especially for supply. Although PACE plans are being developed in the units and at the major commands, no policies exist at the enterprise level for exercising PACE plans.[69] Without exercising PACE plans at the enterprise level, operations in a contested environment are at risk from surprises about data

[68] Communications flyaway kits provide access to NIPRNet, SIPRNet, and Voice over Internet Protocol to a very limited number of users.

[69] An electronic search of all policies in the logistics (20-series), maintenance (21-series), materiel management (23-series), transportation (24-series), logistics staff (25-series), security (31-series), and civil engineering (32-series) functional areas revealed no reference to PACE plans. A similar search of doctrine for logistics also revealed no reference to PACE plans (see Air Force Doctrine Annex 4-0, 2015).

rate and the ability to move information to and from enterprise systems when reverting to alternate, contingency, and emergency communications conduits.

Continuity of Logistics Operations

U.S. Air Force logistics systems are dominantly *pull logistics systems,* which are more fragile in the face of communications and data loss. In the broadest sense, a *push logistics system* is one in which information flows in the same direction as production and supply chains, providing resources based on forecasted demand.[70] These forecasts use past consumption patterns and often include buffer stocks to mitigate potential issues because of incorrect forecasts, such as undersupply.[71] In contrast, a *pull logistics system* is defined by information flows in the opposite direction.[72] Pull systems are driven by signals that downstream destinations are prepared to receive and process material.[73] Although pull systems are beneficial for maximizing efficiency and minimizing excess stocks and transportation costs, they are vulnerable to a breakdown in communications as information flows are needed to trigger the transportation of materials through the supply chain.

Several decades ago, largely to achieve efficiencies, the U.S. Air Force shifted from a dominantly push system to a pull system. Policies and procedures make no provision for moving to a blend of pull and push, or temporarily moving to pure push systems when essential communications with a supported unit are lost. Specifically, who would trigger the switch for supplying a unit is undefined, which organizations would generate the demand signal to the supply system in lieu of the forward unit (e.g., a demand signal to the Defense Logistics Agency) are unspecified, and what planning factors would be used are insufficiently defined.

U.S. Air Force doctrine also does not define the means for devolving authorities to lower echelons and decentralizing command and control via mission-type orders.[74] Mission-type orders have been discussed for decades but not much used.[75]

---

[70] M. C. Bonny, Zongmao Zhang, M. A. Head, C. C. Tien, and R. J. Barson, "Are Push and Pull Systems Really So Different?" *International Journal of Production Economics*, Vol. 59, 1999.

[71] Bhaba R. Sarker and James A. Fitzsimmons, "The Performance of Push and Pull Systems: A Simulation and Comparative Study," *International Journal of Production Research*, Vol. 27, No. 10, 1989.

[72] Bonny et al., 1999.

[73] Michel Baudin, *Lean Logistics: The Nuts and Bolts of Delivering Materials and Goods*, New York: Productivity Press, 2004, p. 35.

[74] A *mission-type order* is "[a]n order to a unit to perform a mission without specifying how it is to be accomplished" (see U.S. Department of Defense, 2020, p. 145).

[75] See, for example, Peterman, 1990.

Summary and Conclusions Regarding the Status Quo

Doctrine for command and control of combat support is designed for a static fight and not one of maneuver. Wings take operational flying orders from the JFACC in the form of an ATO. Attempts to reallocate resources, such as combat support personnel or materiel, from one AEW to another by either the AFFOR staff or a CST within the AOC would likely be met with resistance, as neither of these organizations has the processes and procedures to direct or control combat support resources. Guidance from a CST is not in the form of an order and the ATO does not direct logistics.

The U.S. Air Force lacks processes for issuing a logistics FRAGORD or an organization to issue one. Under the current construct, a logistics FRAGORD to support maneuver would most naturally be issued by the COMAFFOR. The COMAFFOR is at the right vantage point to observe, guide, and direct service efforts within a unified command in a theater. Personnel stationed in the theater may also already be aligned under the COMAFFOR's administrative control. The COMAFFOR also works for the geographic command commander and therefore should be aware, even if not designated as a JFACC, of command priorities. However, given the level of war that the COMAFFOR oversees (operational level, with operational control), this may not be feasible and would lengthen the decision time. Expecting the COMAFFOR to exercise command at all three level of war—strategic, operational, and tactical—for logistics support may demand too much for a single commander.

Planning and executing logistics movement to support maneuver require the coordination of many organizations both inside and outside the theater. Many different data systems and process-control systems are needed to move supplies, equipment, and personnel from one place to another (and for supply, these systems differ across the functional communities). The more coordination that is needed to plan and execute an action, the less agile and timely the outcome.

The decision to maneuver is one that crosses all levels of war. At the tactical level of war, kinetic threats to a base are local and a local commander both has the situational awareness of the dangers and holds responsibility for the personnel at risk. At the operational level of war, maneuver necessarily reallocates resources within the theater, affecting the overall theater posture. And at the strategic level of war, expeditionary basing decisions will often be on the soil of other nation states. Moving assets from one nation state to another or changing the nature of military operations within a nation state would affect the strategic situation and must be handled in concert with diplomatic outreach. Current doctrine does not address the need to distribute the various maneuver decisions to the command authorities at the appropriate levels of war.

In a high-end fight, persistent multi-domain attack would place at risk data integrity, access to data, the ability to communicate, and centralized process control. Command and control processes must be able to adapt to these circumstances to remain effective. They must be able to devolve control gracefully when needed. Existing doctrine and policies do not provide for the explicit delegation of authorities (e.g., through mission-type orders). Policies for establishing relevant planning factors and roles and responsibilities for push logistics also do not exist.

3. Combat Support Command and Control for a High-End Fight

> "Plans are worthless, but planning is everything. There is a very great distinction because when you are planning for an emergency you must start with this one thing: the very definition of 'emergency' is that it is unexpected, therefore it is not going to happen the way you are planning."
>
> —Dwight D. Eisenhower[76]

Some of our findings and possible mitigations in this report are contingent on the operational utility of ground maneuver of air forces. Because the findings and mitigations are predicated on the hypothesis that maneuver is a viable strategy for defending the force and continuing air operations, decisions on investing in one or more of these mitigation strategies should await a fuller demonstration of operational effectiveness of maneuver, a topic beyond the scope of this research effort.

In Chapter One, we identified four themes that run through the objectives for effective command and control of combat support for maneuver and operations while under persistent multi-domain attack:

1. decisionmaking and logistics direction that are timely and distributed when data, communications, or both are degraded
2. situational awareness that is timely and sufficiently comprehensive
3. coordination between combat support and operational activities that is timely and integrated
4. systems and processes for command and control that, when attacked across domains, are robust and resilient.

Comparing these four thematic attributes for combat support with the status quo as outlined in Chapter Two, we arrive at the following central findings.

Findings

A theme running through many of the findings is fragmentation of the command and control of combat support across various functional communities (hereafter, *functional fragmentation*). Functional fragmentation has both positive and negative attributes with respect to the problem of maneuver and logistics operations when under persistent multi-domain attack, attributes that we discuss under multiple headings. The fragmentation is also deeply structural in U.S. Air Force

[76] Public Papers of the Presidents of the United States, "Remarks at the National Defense Executive Reserve Conference," in *Dwight D. Eisenhower, 1957, Containing the Public Messages, Speeches, and Statements of the President, January 1 to December 31, 1957*, Washington, D.C.: Office of the Federal Register Division, National Archives and Records Service, November 14, 1957, p. 818.

logistics processes; it cannot feasibly be changed, and therefore any proposed ways forward need to exploit its benefits and allay its weaknesses.

Decisionmaking

Combat support has no unified command and control mechanisms at the operational level of war. Combat support lies organizationally within the AFFOR staff and is not a central part of the AOC. The AFFOR staff coordinates combat support with many other combat support stakeholders, including the AOC, but does not have accepted and practiced methods to support rapid combat support decisionmaking, such as resource reallocation. Operational control passes from the COMAFFOR directly to the AEW commanders. Expecting the COMAFFOR to exercise command at the strategic, operational, and tactical levels of war for logistics support may be demanding too much of a single commander. This limitation hampers the agility of decisionmaking to support combat support resource reallocation to support maneuver.

The speed of combat support for maneuver is slowed down by the need for the coordinated actions of many actors, both within and outside the theater, to include service, joint, and agency actors. The challenges associated with commanding and controlling logistics capabilities are well documented.[77] Even in steady state, to move personnel, equipment, and supplies—even within the theater—requires the coordinated effort of many different actors, both inside and outside the theater.[78] The more actors involved in decisionmaking, the less timely the process. Maneuver operations in a contested environment may require quick action without assured communications, which would add a level of complexity to disparate processes that are spread across many actors.

When issuing joint orders for the maneuver of air forces to new operating locations, the U.S. Air Force is hampered by its reliance on the ATO for issuing orders. The U.S. Air Force has standardized processes for producing an ATO, which is designed well for its purpose. But in the joint environment, orders for maneuver would more typically be issued as a FRAGORD. Orders in this form are consistent with the practices of the other services. The Department of the Air Force has not established a process for issuing a logistics FRAGORD. Each functional area has a separate process for management of its resources and capabilities.[79]

---

[77] See Tripp et al., 2012. Although the referenced analysis is from 2012, the challenges described in the document persist today. For a detailed description of combat support command and control process gaps and shortfalls, see Lynch, et al., 2014a.

[78] Examples of the fractionation of logistics command and control processes can be found in Lynch, Drew, and Mills, 2018. Also see Kristin F. Lynch, Anthony DeCicco, Bart E. Bennett, John G. Drew, Amanda Kadlec, Vikram Kilambi, Kurt Klein, James Leftwich, Miriam E. Marlier, Ronald G. McGarvey, Patrick H. Mills, Theo Milonopoulos, Robert S. Tripp, and Anna Jean Wirth, *Analysis of Global Management of Air Force War Reserve Materiel to Support Operations in Contested and Degraded Environments*, Santa Monica, Calif.: RAND Corporation, RR-3081-AF, 2021.

[79] Leftwich et al., 2002; Mills et al., 2006; and Lynch et al., 2014a.

The ability to adjust command and control of logistics when under persistent multi-domain attack is impeded by the lack of doctrine, policy, planning, and procedures for distributed command and control and for push logistics. Two major impediments exist: (1) the potential for rapidly changing context for decisionmaking with a pending or actual attack; and (2) a bifurcated logistics decision authority with service and joint command. Distributing command and control when under attack to the lowest levels is hampered by the structural features of combat support systems. Logistics systems cannot be cleanly separated into theater and enterprise components. Centralization and dependence on reachback impair the ability to devolve command and control of combat support within the theater. The lack of planning, procedures, and exercising of push logistics puts at risk the flow of critical resources to units when they lose essential communications.

All of these factors diminish the ability of the combatant commander to exercise his or her directive authority for logistics.

Situational Awareness

The U.S. Air Force has not clearly defined the minimal information needed to maintain situational awareness for maneuver. When maneuver is desired under persistent multi-domain attack, commanders cannot expect full data rate for information flow. They will need to triage data to the minimal needed to support the decisions at hand. The AOC is designed for catered situational awareness of the air and the threats to air operations, including weather. No comparable situational awareness exists for the information needed to inform maneuver situations, and, therefore, no minimal data set has been defined to support decisions on maneuver. The consequence is the risk of not getting the right information to decisionmakers in a crisis or transmitting a surfeit of information that wastes scarce data rate and time.

The U.S. Air Force lacks a common operating picture for combat support. Each functional community has independent situational awareness of the resources within its purview. Information about installations is distributed across various organizations. Logistics information neither flows to a central location nor is placed in a shared environment for common access. For occupied locations, changes in status are reported to the AFFOR staff via situation reports. For unoccupied sites, information is held in surveys, principally base support plans and databases on worldwide airfields. Base support plans are conducted only for sites that have been included in plans and to which the host nation has granted access for that purpose. Although all of the requisite information lies somewhere in the enterprise, specific subsets of the information are not easily available to decisionmakers.

Coordination Between Operations and Combat Support

The separation of the bulk of combat support expertise and personnel (the AFFOR staff) from the locus of operational command and control (the AOC) impedes coordination between operations and combat support. The AFFOR staff performs most of the logistics coordination in a theater. However, because they are a staff, they do not have the authority to direct or control combat support operations and are removed organizationally (and often physically) from the AOC. The AFFOR staff and the AOC staff also often do not use the same computer network enclaves, further hampering information-sharing and coordination. All these factors raise organizational barriers to integrated operations.

The CSTs in AOCs are currently not fully staffed. As of 2019, policy and doctrine still indicate the existence of CSTs in AOCs. At least in their peacetime postures, AOCs in Europe and in the Pacific do not fully staff the CSTs.[80] This posture inhibits logistics coordination within the AOC. Even if the CSTs were stood up and populated during a contingency, the transition and learning during the stand-up would almost certainly curtail its effectiveness during the early period of the conflict. Unlike an AOC, the number of personnel on the AFFOR staffs is based on a nominal staff duty day, with periodic increases during short-term exercises or crisis events. This staffing factor places the AFFOR staff at a disadvantage when working a crisis over an extended period.

The classification of some war plans presents a barrier to the logistics community for adequately planning for supporting operations. A plan is useful only to the degree that the U.S. Air Force can organize, train, and equip to be ready to execute it. Currently, the clearances assigned to some logistics billets are not consistent with the classification of some key parts of war plans that those personnel need to effectively plan support.[81] Either billets need to be revised or key pieces of information (e.g., the number of aircraft and people, operational tempo) that do not need higher classification can be separated and shared with the AFFOR staff while more closely held information, such as location and mission, are not widely shared.

Robustness and Resiliency of Systems and Processes

Functional fragmentation provides some robustness and resiliency to combat support operations. Each functional area has its own supply management systems and processes. That arrangement makes each process and system well-tailored to the unique needs of the function. To alter, delete, or deny access to combat support data across all functional areas, an adversary would have to successfully attack multiple systems. Fragmentation greatly complicates the targeting problem for the adversary.

The U.S. Air Force has limited deployed communications capabilities and capacity for robust and resilient communications. The total number of bases that can be supported with combat communications is limited and probably insufficient to support large-scale maneuver.[82] Maneuver not only increases the number of locations that need support, but the action of maneuver requires a certain quantity of resources for the supply pipeline. The placement of

[80] Discussions with PACAF and USAFE personnel, December 2018.

[81] Discussions with Combat Communications personnel, July 2019.

[82] Discussions with PACAF and USAFE personnel, December 2018.

deployed communications units in the Air Reserve Component requires mobilization for deployment, and therefore takes time, reducing agility.[83] Combat support communications flyaway kits plug into the internet and are therefore not robust against attacks to the internet. Wing-level communications squadrons in the U.S. Air Force have shifted focus over time from generalized communications to providing Internet network services, which diminishes deployed capabilities.

The reliance on enterprise coordination makes operations fragile in a communications degraded environment. A great number of logistics processes require some degree of participation by the centralized enterprise in the form of reachback. For many, a centralized, enterprise information technology system performs process control. Reliance on centralized systems reduces robustness and resiliency.

The U.S. Air Force lacks policy for the prioritization of combat support information in a degraded communications environment. One facet of a degraded communications environment is reduced data rate. When data rate is severely limited, leaders will need to make triage decisions about which data to share, which to delay, and which processes should take precedence. The logistics community does not have policies or plans that establish such priorities. That delays the decisionmaking for prioritization in the time of a crisis.

Several aspects of the current organizational structure and processes for combat support are better suited to peacetime efficiency than wartime effectiveness. Indeed, several were designed with peacetime efficiency in mind to squeeze more capability out of scarce resources. But there is natural tension between peacetime efficiencies and wartime effectiveness. In sum, current combat support command and control processes and authorities are neither optimally designed for the timeliness needed for maneuver nor for the robustness and resiliency for operations while under persistent multi-domain attack.

Potential Mitigation Options for Maneuver

The above findings reveal deficiencies in command and control of logistics, insufficient coordination between operations and combat support, and shortfalls in the situational awareness of the ground picture for maneuver. We present two organizational design options that partially redress these shortfalls, one under the JFACC and the other under the COMAFFOR. We distinguish the two options in the case that the JFACC and COMAFFOR are not the same individual.

Both options aggregate combat support operations into a single organization with command and control authority to issue a logistics FRAGORD. The options differ in the degree to which they explicitly partition the appropriate decisions to commanders at the strategic, operational, and tactical levels of war. A third recommendation, independent of the first two, partially

[83] The 35th Combat Communications Squadron is in the Air Force Reserve Component.

addresses the need for better devolution of command and control when communications are degraded.

Option One: Align Combat Support Command and Control in the AOC

The first option is to assign to the AOC additional responsibilities for combat support and an associated change to the organization of the AOC. This option is depicted in Figure 3.1. Under this option, the AOC would gain the mission of producing orders for logistics and maneuver. In symmetry with the ATO (and the cyber tasking order), it could be called a *logistics tasking order* (LTO), although it could also be issued as a FRAGORD to conform to joint practice.

To accomplish this mission, the AOC would have two directorates. An Air Operations Directorate would replicate the AOC as it is currently constituted, with the five current divisions for generating the master air attack plan and the ATO.[84] A second directorate would be created, which we call the Airfield Operations Directorate, whose mission would be to create the LTO (or logistics FRAGORD). To support the process of generating an LTO, the directorate would need a capability of maintaining a common operating picture for logistics, perhaps in a dedicated division. This division would maintain a common ground picture similar to the common air picture maintained by the status quo AOC. The proposed new directorate would also need to have a plans division and an operations division to generate the LTO. The JFACC would have command authority over tactical maneuver and would exercise that command through the LTO (or logistics FRAGORD), not the ATO. The LTO would be highly coordinated with the ATO, but it would be issued at whatever frequency is needed to support the operations tempo of the theater, in the spirit of a FRAGORD.

---

[84] Air Force Instruction 13-1AOC, 2012.

Figure 3.1. Organizational Design Option One

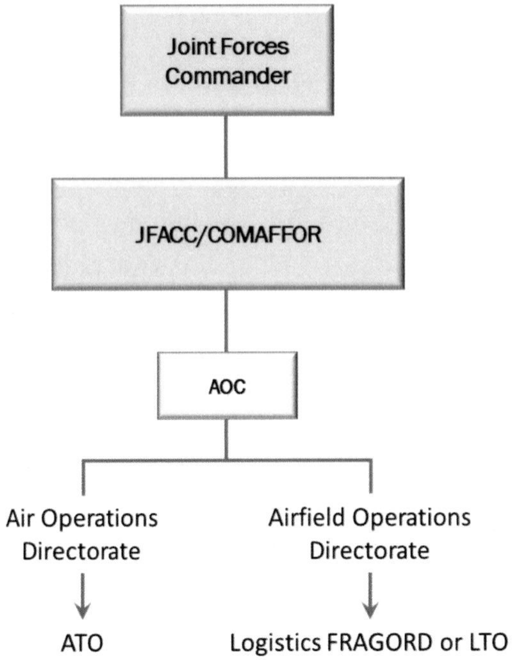

This option has several advantages. By issuing an LTO (or FRAGORD), it provides for combat support operational control at the tactical level of war. This order fosters integration across combat support functions by collating them into a single process and single product. As a directorate in the AOC, the Airfield Operations Directorate would facilitate the integration of combat support and air operations. This integration would include coordinating any military deception supporting maneuver. Situational awareness for combat support would be centralized into a division in the AOC. This structure for combat support command and control would clearly partition staff activities for combat support (assigned to the AFFOR staff) from operational control of combat support (assigned under the JFACC in the AOC, issued as an order).

This option also has several disadvantages. It puts logistics under an operator, outside the immediate control and monitoring of the AFFOR/A4/7 (and the A1 and A6).[85] To mitigate this concern, the head of the Airfield Operations Directorate could be dual-hatted as the AFFOR/A4 when the COMAFFOR and JFACC are the same individual. If the COMAFFOR and the JFACC are not the same individual, however, it would place the organization for the control of U.S. Air

[85] In 2019, both U.S. Air Forces in Europe and Pacific Air Forces reorganized to place the AOC under the AFFOR/A3. The proposed organizational structure in Option One would place the Airfield Operations Directorate in the AOC under the AFFOR/A3.

Force logistics under the command of someone from another service.[86] Even if the directorate were staffed with Department of the Air Force personnel, this situation would seem untenable. Another disadvantage is that it would promote further centralization of processes, making them more fragile when persistently attacked across domains. Devolution structures for distributed command and control and other continuity of operations plans would help ensure robustness and resiliency. A final disadvantage is that it does not provide for an explicit commander at the strategic level of war for maneuver decisions at that level—although, with a change in doctrine, the COMAFFOR could be given that role.

Option Two: Align Combat Support Command and Control Under the COMAFFOR

The second option is to develop organization design elements for the issuance of a logistics FRAGORD under the COMAFFOR, organized under the NEAF/CC (or AETF/CC). This option is depicted in Figure 3.2. Some doctrinal reorganization of the U.S. Air Force service component would be desired under this option. The COMAFFOR in this case would be a three- or four-star general with responsibilities at the strategic and operational levels of war. (Current doctrine only grants the COMAFFOR command at the operational level of war.) The NEAF/CC (or AETF/CC) would be a general of lower rank whose command authority would be at the tactical level of war. Under the command of this officer, an operational organization would be created with the structure of the aforementioned Airfield Operations Directorate. It would have the same proposed structure and produce the same proposed product—the LTO (or FRAGORD)—through the same proposed process as Option One. Although the operational control of combat support falls under the COMAFFOR's command authority, this option still separates operations from staff functions, which would remain with the AFFOR staff.

[86] For example, if the JFC names the naval service component commander as the JFACC, then the Airfield Operations Directorate would be under the command of a naval commander in Option One.

Figure 3.2. Organizational Design Option Two

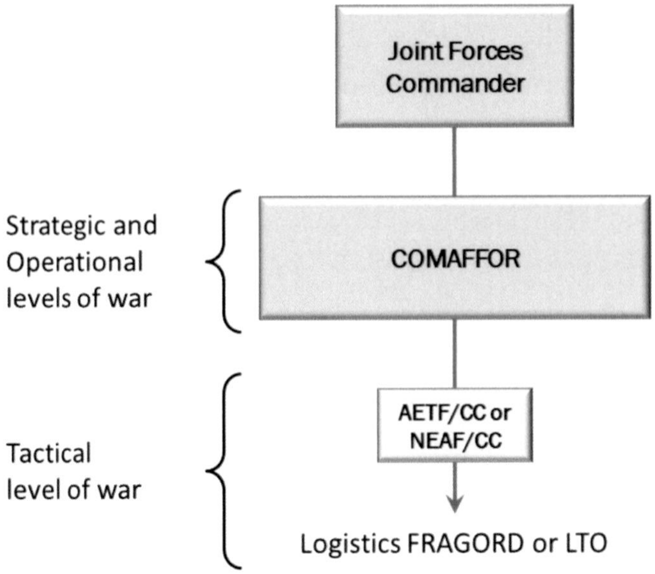

This option has several advantages. This option organizes, trains, and equips the U.S. Air Force for the COMAFFOR to exercise command and control for combat support even when the COMAFFOR and JFACC are not the same individual. It also more cleanly separates command and control at the tactical level from that at the strategic and operational levels of war. Indeed, it allows the COMAFFOR to exercise command at the strategic level, freeing the commander to delegate tactical control to a dedicated commander. This devolution facilitates distributed command and control when needed.

This option also has several disadvantages. The most salient is that it organizationally separates the generation of the ATO from the LTO (logistics FRAGORD), making the desired coordination of these two processes and products more challenging. That separation is also likely to retard the ability to execute maneuver relative to the option that places these two processes under a common command in a single center.

Options One and Two are mutually exclusive. We stay agnostic about a preference for Option One or Two. However, independent of whether one or either is selected, attention must be paid to a separate but important additional issue, the devolution of command and control.

Devolved Command and Control

A key element of distributed command and control is the ability to devolve command to lower echelons. Under persistent multi-domain attack, degraded communications could isolate an installation requiring the personnel to be under some form of local command. Doctrine makes provision for an installation commander but prefers that this commander be an operator who is at least a commander at the squadron echelon level. Doctrine implicitly assumes that the unit is bedded down at the location and that the installation commander will always be physically present. During maneuver, such as that explored in the Combat Support Wing and Rapid Forge exercises, the total number of personnel on the ground could be quite small and the time an operator (pilot) might be on the ground could be short. The location would be used only as a short-term rearming and refueling location, possibly with some limited maintenance capabilities. The operators would be transient and might not include an officer with command at the squadron level or higher. This case differs from the doctrinal case of a beddown location.

To better support a transient maneuver, an option would be for an installation commander to be drawn from the logistics community to oversee BOS-I and operations on the ground. This installation commander would get delegated authorities from the COMAFFOR or NEAF/CC under mission-type orders and, when so delegated, make decisions about when to move operations out of the location. Operators transiting through the installation would remain under the command of the JFACC and follow the ATO.

Drawing a commander from the logistics community to be an installation commander with potential authorities to move operations, when so delegated by mission-type orders, would be a new responsibility for combat support officers. If such a course of action were chosen, it would drive the need for military decisionmaking processes to be taught much earlier in logistics officers' careers to prepare them for this role, akin to how Army officers receive early training in command. We recommend that such steps begin.

General Recommendation Related to Maneuver

A general recommendation for maneuver is to develop preliminary base support plans for all potential locations that might be used for maneuver. Under current policy, base support plans are drawn up only for sites that explicitly appear in approved war plans. When plans are made, the fidelity and currency of these plans are sometimes limited by the ability of survey teams to access the locations. Maneuver requires certain information about a location prior to use to accurately judge whether to use the site and what additional infrastructure and support are needed. The more information that is gleaned in advance about potential sites, the higher degree to which this information is kept up to date, and the higher degree to which this information is made available across the enterprise, the less time will be consumed by planning and preparation and the more likely better decisions will be made for maneuver during a crisis.

The need for site information continues into the time of operations. There is a need to define minimal information about each potential maneuver site that a commander would need to know for situational awareness when making a maneuver decision.

A list of minimal essential information about sites could be assembled that would cover more locations than base support plans and not be as onerous to assemble. Potential information for such a database could be

- airfield information, including a line drawn map of the airfield with attention to runway length and width, and available parking areas (This information can be compiled with available information sources without the need for a site survey team.)

- fuel information, including available storage capacity, resupply capacity, and whether hydrants or trucks are available to refuel
- available infrastructure, including infrastructure for
  - base support (e.g., billeting, potable water, food sources, hospitals, fire protection, and vehicle availability)
  - sortie generation (e.g., hangers, shops, engine hush house or trim pad, munitions build-up and storage sites).
- utilization plans, including what other organizations plan to use the installation, for how long, and doing what
- host-nation–provided capability, including what airfield instrumentation exists, communications capabilities, arresting system availability, and any other host-nation–provided capabilities that the U.S. Air Force could use or augment.

For bases in use during a contingency, the minimal essential information list for a situation report that could be manually carried out of the site by a pilot could be defined. This information could be organized in the following fields:

- **Sorties flown:** Typically, this would be what was flown the day prior or at the end of the flying day and could be represented by simply 12X12X12 with average sortie duration of 2.5, which would represent three goes of 12 aircraft with each airplane flying 2.5 hours per sortie. From this, fuel and scheduled maintenance for both engines and aircraft could be estimated. If the rear unit knows the standard configuration listing (weapons loading plan) that was loaded on each aircraft, the unit could also predict weapon expenditure rates.
- **Days of supply and pacing item:** The unit could subdivide this area into fuels, munitions (both aircraft and small arms), water, food, or any other high-interest items. The actual information transmitted could be represented by a number or a fraction. Petroleum, oil, and lubricants 2.5 would indicate 2.5 days of flying without a resupply of fuel. The unit could also list any highly problematic parts in the area.
- **Status of personnel:** In this area, the unit could list wounded in action, killed in action, and missing in action, and highlight any critical skills significantly negatively affecting operations.
- **Status of infrastructure:** The unit could subdivide this area into sortie generation and sortie support.
  - In sortie generation, the unit would report on runway, ramps, airfield lighting, airfield instrumentation, arresting systems, tower, maintenance facilities, and general communication issues.
  - In sortie support area, the unit could report on housing, feeding, shower and shave, fire protection, security forces, vehicle maintenance, and administrative infrastructure needs or damage. (For reports with no damage, the report could be as simple as sortie generation green and sortie support green and the unit would only list exemptions.)
- **Threat environment:** The unit could report tactical threat information regarding risks of which forces coming to the base should be aware.

- **Commanders assessment and problem issues:** This field could permit a short narrative where the commander can describe any issues or problem areas where the unit needs help or things perceived as possible imminent issues.

Current policy supports the development and use of base support plans and minimum essential information, including the information needed in a situation report. The U.S. Air Force needs to update tactics, techniques, and procedures to conduct and develop these plans and to gather the information. It would need to fund the requirement to complete these activities. All three of the organizational options presented in this chapter would require changes to doctrine, and all are predicated on the operational effectiveness of the maneuver the options are meant to facilitate.

Further Recommendations

The previous options are directed toward changing command relations, processes, and organizations to better support the temporal demands of maneuver. The following recommendations are unconditional with respect to operational concepts. They promote general expeditionary readiness and support robustness and resiliency of combat support command and control under persistent multi-domain attack. Cost considerations are the main extenuating condition over their adoption.

The first recommendation is a broad one that touches on many challenges—to ensure that the right personnel have access to the information that they need to do their jobs.

To ensure that the appropriate Airmen have access to the planning information that they need, use a combination of (1) extracting lower classification material from war plans and (2) assigning a higher classification to key logistics billets.

Diverse Communications

Ensure that the logistics community has multiple, diverse means to communicate and control processes, and that these are regularly exercised. Policy already provides for PACE plans but not at the enterprise level. The PACE construct provides for multiple communications pathways. But these pathways must be useful for each logistics demand. To be ready for persistent multi-domain attack, the modes must have low probability of correlated failure and must be routinely exercised. Logistics in a theater cannot be cleanly separated from the greater enterprise, so PACE plans should be enterprisewide. The entire logistics enterprise becomes warfighting support during a high-end conflict, so the entire enterprise needs to be part of PACE plans, not just the service component to the combatant commander. Peacetime logistics processes should use alternate, contingency, and emergency pathways routinely. These exercises should inform improvements to the PACE plan and prepare personnel for the use of the full spectrum of communications pathways when under persistent multi-domain attack.

Robust and Resilient Data

Instruct local units to regularly back up data critical to their needs and inspect that they do so.[87] Attacks can lead to loss of access to data, data deletion, and corruption of data. Centralized solutions for backing up data are costly and more fragile than local ones. To back up a large, enterprise system like the Standard Base Supply System entails multiple data centers—and if those are merely homogeneously redundant, central backup still leaves the system exposed to common attack vectors. Even diverse centers still present adversaries with a few targets. Yet these enterprise systems are often the repositories of record for data needed for local operations, down to the specific shelf locations for supplies. Rather than back up the entire system centrally, local units should, on a regular basis, back up the small fraction of those data that are critical for their operations. For example, a unit running a supply warehouse could daily burn to a compact disc the data for the locations of the supplies in its warehouse. Upon corruption or loss of access to the enterprise data, the unit could revert to the last known reliable state.

The frequency and scope of these backups could be adjusted to the threat environment. If the threat is low, the backups could be conducted at less-frequent intervals and be less comprehensive. When the threat is higher during a contingency, the intervals between backups should be shorter and more comprehensive. The logistics FRAGORD could direct the frequency of backups and key the conditions to the Information Operations Conditions system levels. Note that policy permits Information Operations Condition system levels to be set more strictly locally by the relevant commander.[88]

Command and Control for Push Logistics

To mitigate against the loss of supporting information systems, develop and exercise the ability to shift from pull to push logistics.

The ability to switch temporarily to a push logistics system for a unit requires several elements of command and control.

First, assign responsibilities for who has command authority to delegate the ability to generate a demand signal for supplies in lieu of the forward unit. Even in a push system, some agent must generate a demand signal. In a pull system, that agent is the forward unit at the consuming end of the supply chain. In a push system, it is some organization upstream. Doctrine would need to assign authorities for when to empower an organization upstream to take over the generation of the demand signal. The COMAFFOR, through Title 10 service authorities, may be well positioned doctrinally for this authority if the combatant commander delegates directive authority for logistics.

---

[87] Snyder et al., 2017.

[88] To become the Cyber Conditions system. Chairman of the Joint Chiefs of Staff Instruction 6510.01F, *Information Assurance (IA) and Support to Computer Network Defense (CND)*, Washington, D.C., June 9, 2015.

Second, assign responsibilities for which organization(s) will generate that demand signal. If communications allow, the AFFOR staff may be well positioned to take over the generation of the demand signal; however, the AFFOR staff might not have access to all the tools and systems needed to generate requirements (e.g., the Defense Logistics Agency may be better positioned to generate demand for fuel). Provision should be made for procedures if the AFFOR staff are also victims of communications loss and cannot carry out these responsibilities. In that case, a possibility would be to delegate the generation of the demand signal to a crisis action team at Air Combat Command because of its global force management responsibilities, or combat support global managers, such as the Global Ammunition Control Point for munitions.

Third, establish process and planning factors to generate the demand signal. The first two elements are adjustments to doctrine and policy. The third requires assembling existing data in the U.S. Air Force, gathering additional data, and developing appropriate processes.

We envision a process for push logistics with three strands: (1) situational awareness of the receiving unit's circumstances; (2) information regarding transportation and storage availability; and (3) planning factors to estimate additional resources the unit is likely to need.

The first strand is critical information on the immediate situation of the receiving unit. This information would include the quantity of supply and pacing of items at the base, the capability of the base to receive resources, the status of personnel, and the status of equipment for sortie generation and sortie support, because these components will determine the type of planning factors (described next) that would be needed to send materiel or equipment.[89] This could also include such information as the availability of pre-positioned material, host-nation support, or the ability to procure supplies on the local market.

The second strand is critical information regarding transportation modes and storage capabilities. There is no point in attempting to send resources to a unit if they cannot reach the destination or if there is no ability to receive and store the resources when they arrive.

The third strand is planning factors. The underlying data and modeling tools needed for such planning factors largely exist, scattered throughout the U.S. Air Force. Here, we describe several examples from different functional communities. For repair parts, the flying hour program tools to predict repair part needs can be combined with the ATO and War Mobilization Plan Volume Five factors to anticipate repair part needs. Other supporting tools include the Aircraft Availability Model, which provides cost and availability rates for various types of aircraft; the Aircraft Sustainability Model, which defines a spare parts kit that units will take for a given period of time; and the Customer-Oriented Leveling Technique model, which recommends adjustments to repair part allocation.[90] However, contractor logistics support for some aircraft complicates the ability of the U.S. Air Force to do push logistics organically.

---

[89] This also includes critical skills and information on runways, ramps, lighting, instrumentation, and other factors.

[90] T. J. O'Malley, *The Aircraft Availability Model: Conceptual Framework and Mathematics*, Washington, D.C.: Logistics Management Institute, June 1983; F. Michael Slay, Tovey C. Bachman, Robert C. Kline, T. J. O'Malley,

Munitions could be estimated using Global Ammunition Control Point systems. These systems include planning factors for aircraft type, standard configured load, and expenditure rates. The Combat Forces Ammunition Model matches weapons to targets. Perhaps this model could be run in reverse to estimate munitions based on targets. For rations, the number of personnel drives the requirement, and planning factors exist for these within the Air Force Services Agency of AFIMSC. Carefully crafted base support plans would identify resources and capabilities at a forward operating location, and any limiting factors or shortfalls that could be used to calculate needs at a base after an attack.[91] For equipment, prototype models could be used to calculate needs if the U.S. Air Force were to adopt them.[92] Table 3.1 summarizes these and some additional planning factors for several categories of support.

These data need to be brought together in advance of the need. A paradigm for push logistics that relies on connectivity and information-sharing among a variety of organizations, databases, and tools to assemble the planning factors at wartime is incompatible with organizing, training, and equipping for a persistent multi-domain attack.

Frank L. Eichorn, and Randall M. King, *Optimizing Spares Support: The Aircraft Sustainability Model*, Washington, D.C.: Logistics Management Institute, October 1996; and Jason Vinson and Kevin Gaudette, "Customer-Oriented Leveling Technique," *Air Force Journal of Logistics*, Vol. 27, No. 1, Spring 2003, pp. 16–19.

[91] Air Force Instruction 10-404, *Base Support and Expeditionary (BAS&E) Site Planning*, Washington, D.C.: Secretary of the Air Force, July 24, 2019, p. 9.

[92] Snyder and Mills, 2004; and Patrick Mills, James A. Leftwich, Kristin Van Abel, and Jason Mastbaum, *Estimating Air Force Deployment Requirements for Lean Force Packages: A Methodology and Decision Support Tool Prototype*, Santa Monica, Calif.: RAND Corporation, RR-1855-AF, 2017.

Table 3.1. Summary of Information Needed to Develop a Logistics Push Package

Category	Functional Areas	Organizations	Planning Factors	Additional Elements	Forecasting Models
Rations	Force support	Air Force Services Agency, AFFOR/A1	• Food: Meals per person per day • Water: Gallons per person per day	• Equipment • Storage	• Planning guides
Fuel	Logistics readiness	AFPET, AFFOR/A4, DLA Energy	• Aircraft type • Sortie rate and duration • Reserve factor	• Equipment • Storage • Additives	• Fuels support equipment calculator • DLA Energy fuel consumption
Spare parts	Logistics readiness, maintenance	Supply Chain Operations Wing, DLA Aviation, AFFOR/A4	• Mission-design series • Sortie rate and duration	• Manpower requirements	• Aircraft Availability Model • Aircraft Sustainability Model • COLT
Munitions	Munitions management, materiel management	Global Ammunition Control Point	• Aircraft type • Standard configured load • Expenditure rates	• Assembly • Storage • Loaders	• Combat Forces Ammunition Model

NOTE: AFPET = Air Force Petroleum Office; COLT = Customer-Oriented Leveling Technique; DLA = Defense Logistics Agency.

The Role of AFIMSC

The centralization of the assigned combat support planning and budgeting activities to AFIMSC supports the goal of peacetime efficiencies.[93] And by centralizing expertise in various functional areas, AFIMSC can also support the warfighter during operations, including such a time-critical activity as maneuver. AFIMSC could serve as the single interface between the theater AFFOR staff and the functional support for services, security forces, and civil engineering. The Expeditionary Support Directorate could enhance its reachback support capabilities for the warfighter by establishing a permanent cell. The cell would not need to solve all queries from the theaters but would be able to direct the AFFOR staff to the right resources. In other words, the cell could serve as a kind of a clearinghouse for directing AFFOR staff to the

[93] U.S. Air Force, *Implementation of the Air Force Installation and Mission Support Center (AFIMSC)*, Washington, D.C.: Headquarters U.S. Air Force, Program Action Directive (PAD) 14-04, February 25, 2015, Not available to the general public.

right reachback support in the greater enterprise.[94] To be most effective, during stressing operations the support cell would need to be staffed around the clock every day. A reachback capability is only as good as the survivability of its communications links, so robust and resilient communications links between the AFIMSC expeditionary operations cell and forward AFFOR staffs would be critical to its success. A peacetime staff would need to be trained and ready to fill this role, including exercising all contingency communications modes.

Conclusions

A high-end fight places heavy demands on the command and control of combat support. Relative to operations in the past three decades, there is a greater need for decisionmaking that is faster. Such decisionmaking demands better situational awareness on the ground, increased coordination between operators and logisticians, and processes that are more robust and resilient. Our principal recommendations to meet these needs are to

- establish some authority within the theater who can issue a FRAGORD for logistics (We presented options for placing such an authority under the COMAFFOR and the JFACC.)
- define doctrinal means by which command and control can devolve to the base level
- define minimal essential information for locations, both occupied and not occupied by the U.S. Air Force, for situational awareness to support decisionmaking for maneuver
- ensure that Airmen have access to the war planning information that they need to perform the logistics planning
- ensure multiple enterprisewide robust communications pathways for logistics and exercise their use
- define and follow procedures for backing up critical logistics data locally
- develop and exercise procedures for the command and control of push logistics
- establish the ability to provide an operational support cell 24 hours per day, seven days per week within AFIMSC to support the warfighter during wartime.

Although these recommendations will not solve all of the challenges of the command and control of combat support in a high-end fight, few are expensive to implement, because they are process (including organizational) and doctrine changes, not material, and should bring benefits to the warfighter in any conflict.

---

[94] This concept is similar to a logistics response cell set up within Air Force Materiel Command during Operation Enduring Freedom (Afghanistan).

References

5th Combat Communications Group, *5 CCG Planners Guide*, May 30, 2018.

Air Force Doctrine Annex 3-30, *Command and Control*, Maxwell Air Force Base, Ala.: Curtis E. LeMay Center for Doctrine and Education, November 7, 2014.

Air Force Doctrine Annex 4-0, *Combat Support*, Maxwell Air Force Base, Ala.: Curtis E. LeMay Center for Doctrine and Education, December 21, 2015.

Air Force Instruction 10-404, *Base Support and Expeditionary (BAS&E) Site Planning*, Washington, D.C.: Secretary of the Air Force, July 24, 2019.

Air Force Instruction 13-103, *AFFOR Staff Operations, Readiness, and Structures*, Washington, D.C.: Secretary of the Air Force, August 19, 2014.

Air Force Instruction 13-1AOC, *Operational Procedures: Air Operations Center (AOC)*, Washington, D.C.: Secretary of the Air Force, November 2, 2011, incorporating change 1, May 18, 2012.

Amouzegar, Mahyar A., Robert S. Tripp, Ronald G. McGarvey, Edward W. Chan, and C. Robert Roll, Jr., *Supporting Air and Space Expeditionary Forces: Analysis of Combat Support Basing Options*, Santa Monica, Calif.: RAND Corporation, MG-261-AF, 2004. As of July 2, 2020:
https://www.rand.org/pubs/monographs/MG261.html

Baudin, Michel, *Lean Logistics: The Nuts and Bolts of Delivering Materials and Goods*, New York: Productivity Press, 2004.

Blair, Bruce G., *Strategic Command and Control: Refining the Nuclear Threat*, Washington, D.C.: Brookings Institution, 1985.

Bonny, M. C., Zongmao Zhang, M. A. Head, C. C. Tien, and R. J. Barson, "Are Push and Pull Systems Really So Different?" *International Journal of Production Economics*, Vol. 59, 1999, pp. 53–64.

Briggs, Michael, "Air Force Completes Test of Combat Support Innovation," Air Force News Service, May 14, 2019. As of May 16, 2019:
https://www.af.mil/News/Article-Display/Article/1847334/air-force-completes-test-of-combat-support-innovation/

Bruckbauer, Brian, Kevin Heckle, and Janelle Galang, *Combat Support Wing Capstone Exercise After-Action Report*, Joint Base San Antonio Lackland, Tex.: Air Force Installation and Mission Support Center, June 10, 2019, Not available to the general public.

Chairman of the Joint Chiefs of Staff Instruction 4120.02D, *List of Priorities—DoD Transportation Movement Priority System*, Washington, D.C., May 3, 2019.

Chairman of the Joint Chiefs of Staff Instruction 6510.01F, *Information Assurance (IA) and Support to Computer Network Defense (CND)*, Washington, D.C., June 9, 2015.

Cope, Kyle, "Operation Rapid Forge Concludes," Air Force News Service, July 28, 2019. As of July 29, 2019: https://www.af.mil/News/Article-Display/Article/1918143/operation-rapid-forge-concludes/

Feinberg, Amatzia, Hyman L. Shulman, Louis W. Miller, and Robert S. Tripp, *Supporting Expeditionary Aerospace Forces: Expanded Analysis of LANTIRN [Low Altitude Navigation and Targeting Infrared for Night] Options*, Santa Monica, Calif.: RAND Corporation, MR-1225-AF, 2001. As of July 2, 2020: https://www.rand.org/pubs/monograph_reports/MR1225.html

Feinberg, Amatzia, Eric Peltz, James A. Leftwich, Robert S. Tripp, Mahyar A. Amouzegar, Russell Grunch, John G. Drew, Tom LaTourrette, and Charles Robert Roll, Jr., *Supporting Expeditionary Aerospace Forces: Lessons from the Air War over Serbia*, Santa Monica, Calif.: RAND Corporation, 2002, Not available to the general public.

Freedman, Lawrence, *Strategy: A History*, New York: Oxford University Press, 2013.

Galway, Lionel, Robert S. Tripp, Timothy L. Ramey, and John G. Drew, *Supporting Expeditionary Aerospace Forces: New Agile Combat Support Postures*, Santa Monica, Calif.: RAND Corporation, MR-1075-AF, 2000. As of July 2, 2020: https://www.rand.org/pubs/monograph_reports/MR1075.html

Garbarino, Micah, "Hill AFB Airmen Expand F-35A Combat Capability in Rapid Forge," Air Force News Service, July 19, 2019. As of July 29, 2019: https://www.af.mil/News/Article-Display/Article/1910723/hill-afb-airmen-expand-f-35a-combat-capability-in-rapid-forge/

Greenberg, Andy, "The Code That Crashed the World: The Untold Story of NotPetya, the Most Devastating Cyberattack in History," *Wired*, September 2018, pp. 52–63. As of April 29, 2019: https://www.wired.com/story/notpetya-cyberattack-ukraine-russia-code-crashed-the-world

Jablonsky, David, *War by Land, Sea, and Air: Dwight Eisenhower and the Concept of Unified Command*, New Haven, Conn.: Yale University Press, 2010.

Joint Publication 1, *Doctrine for the Armed Forces of the United States*, Washington, D.C.: Joint Chiefs of Staff, incorporating change 1, July 12, 2017.

Joint Publication 3-0, *Joint Operations*, Washington, D.C.: Joint Chiefs of Staff, incorporating change 1, October 22, 2018.

Joint Publication 3-30, *Joint Air Operations*, Washington, D.C.: Joint Chiefs of Staff, July 25, 2019.

Joint Publication 4-0, *Joint Logistics*, Washington, D.C.: Joint Chiefs of Staff, incorporating change 1, May 8, 2019.

Joint Publication 4-04, *Contingency Basing*, Washington, D.C.: Joint Chiefs of Staff, January 4, 2019.

Joint Publication 4-1, *The Defense Transportation System*, Washington, D.C.: Joint Chiefs of Staff, July 18, 2017.

Joint Publication 5-0, *Joint Planning*, Washington, D.C.: Joint Chiefs of Staff, June 16, 2017.

Kasischke, Staci, "Agile Lightning Demonstrates Nimble Operations," U.S. Air Forces Central Command Public Affairs, August 12, 2019. As of August 14, 2019: https://www.afcent.af.mil/News/Article/1929238/agile-lightning-demonstrates-nimble-operations/

Leftwich, James, Robert Tripp, Amanda Geller, Patrick Mills, Tom LaTourrette, C. Robert Roll, Jr., Cauley von Hoffman, and David Johansen, *Supporting Expeditionary Aerospace Forces: An Operational Architecture for Combat Support Execution Planning and Control*, Santa Monica, Calif.: RAND Corporation, MR-1536-AF, 2002. As of July 2, 2020: https://www.rand.org/pubs/monograph_reports/MR1536.html

Losey, Stephen, "Cops Refueling Jets: Small Teams, Sharing Jobs, Train to Set Up Airfields in Combat," *Air Force Times*, May 15, 2019.

Lynch, Kristin F., John G. Drew, Robert S. Tripp, and C. Robert Roll, Jr., *Supporting Air and Space Expeditionary Forces: Lessons from Operation Iraqi Freedom*, Santa Monica, Calif.: RAND Corporation, MG-193-AF, 2005. As of July 2, 2020: https://www.rand.org/pubs/monographs/MG193.html

Lynch, Kristin F., and William A. Williams, *Combat Support Execution Planning and Control: An Assessment of Initial Implementations in Air Force Exercises*, Santa Monica, Calif.: RAND Corporation, TR-356-AF, 2009. As of June 22, 2020: https://www.rand.org/pubs/technical_reports/TR356.html

Lynch, Kristin F., John G. Drew, Amy L. Maletic, Robert S. Tripp, Ricardo Sanchez, William A. Williams, Brent Thomas, and Max Woodworth, *A Strategic Assessment of Component Numbered Air Force (C-NAF) Force Postures*, Santa Monica, Calif.: RAND Corporation, 2010, Not available to the general public.

Lynch, Kristin F., John G. Drew, Robert S. Tripp, Daniel M. Romano, Jin Woo Yi, and Amy L. Maletic, *An Operational Architecture for Improving Air Force Command and Control Through Enhanced Agile Combat Support Planning, Execution, Monitoring, and Control*

Processes, Santa Monica, Calif.: RAND Corporation, RR-261-AF, 2014a. As of July 2, 2020:
https://www.rand.org/pubs/research_reports/RR261.html

Lynch, Kristin F., John G. Drew, Robert S. Tripp, Daniel M. Romano, Jin Woo Yi, and Amy L. Maletic, *Implementation Actions for Improving Air Force Command and Control Through Enhanced Agile Combat Support Planning, Execution, Monitoring, and Control Processes*, Santa Monica, Calif.: RAND Corporation, RR-259-AF, 2014b. As of July 2, 2020: https://www.rand.org/pubs/research_reports/RR259.html

Lynch, Kristin F., John G. Drew, and Patrick Mills, *Enhancing Air Force Materiel Command Support to the Warfighter*, Santa Monica, Calif.: RAND Corporation, RR-2255-AF, 2018. As of June 22, 2020:
https://www.rand.org/pubs/research_reports/RR2255.html

Lynch, Kristin F., Anthony DeCicco, Bart E. Bennett, John G. Drew, Amanda Kadlec, Vikram Kilambi, Kurt Klein, James Leftwich, Miriam E. Marlier, Ronald G. McGarvey, Patrick H. Mills, Theo Milonopoulos, Robert S. Tripp, and Anna Jean Wirth, *Analysis of Global Management of Air Force War Reserve Materiel to Support Operations in Contested and Degraded Environments*, Santa Monica, Calif.: RAND Corporation, RR-3081-AF, 2021. As of February 12, 2021:
https://www.rand.org/pubs/research_reports/RR3081.html

McGarvey, Ronald G., James M. Masters, Louis Luangkesorn, Stephen Sheehy, John G. Drew, Robert Kerchner, Ben D. Van Roo, and Charles Robert Roll, Jr., *Supporting Air and Space Expeditionary Forces [AEFs]: Analysis of CONUS [Continental U.S.] Centralized Intermediate Repair Facilities*, Santa Monica, Calif.: RAND Corporation, MG-418-AF, 2008. As of June 22, 2020:
https://www.rand.org/pubs/monographs/MG418.html

McGarvey, Ronald G., Robert S. Tripp, Rachel Rue, Thomas Lang, Jerry M. Sollinger, Whitney A. Conner, and Louis Luangkesorn, *Global Combat Support Basing: Robust Prepositioning Strategies for Air Force War Reserve Materiel [WRM]*, Santa Monica, Calif.: RAND Corporation, MG-902-AF, 2010. As of June 22, 2020:
https://www.rand.org/pubs/monographs/MG902.html

Mills, Patrick, Ken Evers, Donna Kinlin, and Robert S. Tripp, *Supporting Air and Space Expeditionary Forces: Expanded Operational Architecture for Combat Support Execution Planning and Control*, Santa Monica, Calif.: RAND Corporation, MG-316-AF, 2006. As of July 1, 2020:
https://www.rand.org/pubs/monographs/MG316.html

Mills, Patrick, John G. Drew, John A. Ausink, Daniel M. Romano, and Rachel Costello, *Balancing Agile Combat Support Manpower to Better Meet the Future Security Environment*, Santa Monica, Calif.: RAND Corporation, RR-337-AF, 2014. As of June 22, 2020: https://www.rand.org/pubs/research_reports/RR337.html

Mills, Patrick, James A. Leftwich, Kristin Van Abel, and Jason Mastbaum, *Estimating Air Force Deployment Requirements for Lean Force Packages: A Methodology and Decision Support Tool Prototype*, Santa Monica, Calif.: RAND Corporation, RR-1855-AF, 2017. As of June 22, 2020: https://www.rand.org/pubs/research_reports/RR1855.html

Mueller, Karl P., ed., *Precision and Purpose: Airpower in the Libyan Civil War*, Santa Monica, Calif.: RAND Corporation, RR-676-AF, 2015. As of June 22, 2020: https://www.rand.org/pubs/research_reports/RR676.html

O'Malley, T.J., *The Aircraft Availability Model: Conceptual Framework and Mathematics*, Washington, D.C.: Logistics Management Institute, June 1983.

Osborne, Tony, "Sweden Is Relearning Cold War Basing Skills," *Aviation Week & Space Technology*, July 29–August 18, 2019, p. 20.

Peltz, Eric, Hyman L. Shulman, Robert S. Tripp, Timothy L. Ramey, and John G. Drew, *Supporting Expeditionary Aerospace Forces: An Analysis of F-15 Avionics Options*, Santa Monica, Calif.: RAND Corporation, MR-1174-AF, 2000. As of July 2, 2020: https://www.rand.org/pubs/monograph_reports/MR1174.html

Perrow, Charles, *The Next Catastrophe: Reducing Our Vulnerabilities to Natural, Industrial, and Terrorist Disasters*, Princeton, N.J.: Princeton University Press, 2007.

Peterman, Robert W., *Mission-Type Orders: An Employment Concept for the Future*, Maxwell Air Force Base, Ala.: Air University, Air War College Research Report, March 1990.

Priebe, Miranda, Alan J. Vick, Jacob L. Heim, and Meagan L. Smith, *Distributed Operations in a Contested Environment: Implications for USAF Force Presentation*, Santa Monica, Calif.: RAND Corporation, RR-2959-AF, 2019. As of June 22, 2020: https://www.rand.org/pubs/research_reports/RR2959.html

Public Papers of the Presidents of the United States, "Remarks at the National Defense Executive Reserve Conference," in *Dwight D. Eisenhower, 1957, Containing the Public Messages, Speeches, and Statements of the President, January 1 to December 31, 1957*, Washington, D.C.: Office of the Federal Register Division, National Archives and Records Service, November 14, 1957. As of July 1, 2020: https://quod.lib.umich.edu/p/ppotpus/4728417.1957.001/37

Sarker, Bhaba R., and James A. Fitzsimmons, "The Performance of Push and Pull Systems: A Simulation and Comparative Study," *International Journal of Production Research*, Vol. 27, No. 10, 1989, pp. 1715–1731.

Schanz, Marc V., "Rapid Raptors: A New PACAF Concept Gets F-22s to the Fight Fast," *Air Force Magazine*, November 2013, p. 57.

Slay, F. Michael, Tovey C. Bachman, Robert C. Kline, T. J. O'Malley, Frank L. Eichorn, and Randall M. King, *Optimizing Spares Support: The Aircraft Sustainability Model*, Washington, D.C.: Logistics Management Institute, October 1996.

Snyder, Don, and Patrick Mills, *Supporting Air and Space Expeditionary Forces: A Methodology for Determining Air Force Deployment Requirements*, Santa Monica, Calif.: RAND Corporation, MG-176-AF, 2004. As of June 22, 2020: https://www.rand.org/pubs/monographs/MG176.html

Snyder, Don, George E. Hart, Kristin F. Lynch, and John G. Drew, *Ensuring U.S. Air Force Operations During Cyber Attacks Against Combat Support Systems: Guidance for Where to Focus Mitigation Efforts*, Santa Monica, Calif.: RAND Corporation, RR-620-AF, 2015. As of June 22, 2020: https://www.rand.org/pubs/research_reports/RR620.html

Snyder, Don, Elizabeth Bodine-Baron, Mahyar A. Amouzegar, Kristin F. Lynch, Mary Lee, and John G. Drew, *Robust and Resilient Logistics Operations in a Degraded Information Environment*, Santa Monica, Calif.: RAND Corporation, RR-2015-AF, 2017. As of June 22, 2020: https://www.rand.org/pubs/research_reports/RR2015.html

"Tonga Hit By Near-Total Internet Blackout," BBC News, January 23, 2019. As of April 29, 2019: https://www.bbc.co.uk/news/world-asia-46968752

Tripp, Robert S., Lionel Galway, Paul Killingsworth, Eric Peltz, Timothy L. Ramey, and John G. Drew, *Supporting Expeditionary Aerospace Forces: An Integrated Strategic Agile Combat Support Planning Framework*, Santa Monica, Calif.: RAND Corporation, MR-1056-AF, 1999. As of July 1, 2020: https://www.rand.org/pubs/monograph_reports/MR1056.html

Tripp, Robert S., Lionel A. Galway, Timothy L. Ramey, Mahyar A. Amouzegar, and Eric Peltz, *Supporting Expeditionary Aerospace Forces: A Concept for Evolving the Agile Combat Support/Mobility System of the Future*, Santa Monica, Calif.: RAND Corporation, MR-1179-AF, 2000. As of July 1, 2020: https://www.rand.org/pubs/monograph_reports/MR1179.html

Tripp, Robert S., Kristin F. Lynch, John G. Drew, and Edward W. Chan, *Supporting Air and Space Expeditionary Forces: Lessons from Operation Enduring Freedom*, Santa Monica, Calif.: RAND Corporation, MR-1819-AF, 2004. As of September 29, 2020: https://www.rand.org/pubs/monograph_reports/MR1819.html

Tripp, Robert S., Kristin F. Lynch, Charles Robert Roll, Jr., John G. Drew, and Patrick Mills, *A Framework for Enhancing Airlift Planning and Execution Capabilities Within the Joint Expeditionary Movement System*, Santa Monica, Calif.: RAND Corporation, MG-377-AF, 2006. As of July 1, 2020: https://www.rand.org/pubs/monographs/MG377.html

Tripp, Robert S., William A. Williams, Kristin F. Lynch, John G. Drew, Dahlia S. Lichter, and Laura H. Baldwin, *A Strategic Analysis of Air and Space Operations Center Force Posture Options*, Santa Monica, Calif.: RAND Corporation, 2008, Not available to the general public.

Tripp, Robert S., Kristin F. Lynch, John G. Drew, and Robert G. DeFeo, *Improving Air Force Command and Control Through Enhanced Agile Combat Support Planning, Execution, Monitoring, and Control Processes*, Santa Monica, Calif.: RAND Corporation, MG-1070-AF, 2012. As of June 22, 2020: https://www.rand.org/pubs/monographs/MG1070.html

U.S. Air Force, *Implementation of the Air Force Installation and Mission Support Center (AFIMSC)*, Washington, D.C.: Headquarters U.S. Air Force, Program Action Directive (PAD) 14-04, February 25, 2015, Not available to the general public.

U.S. Code, Title 10, Section 164, Commanders of Combatant Commands: Assignment; Powers and Duties.

U.S. Department of Defense, *Summary of the 2018 National Defense Strategy of the United States of America: Sharpening the American Military's Competitive Edge*, Washington, D.C., 2018.

U.S. Department of Defense, *DOD Dictionary of Military and Associated Terms*, Washington, D.C., January 2020.

Vinson, Jason, and Kevin Gaudette, "Customer-Oriented Leveling Technique," *Air Force Journal of Logistics*, Vol. 27, No. 1, Spring 2003, pp. 16–19.

Wellock, Michael M., "Getting Out: Securing Air Bases During a Withdrawal," in Col Shannon W. Caudill, ed., *Defending Air Bases in an Age of Insurgency*, Vol. 2, Maxwell Air Force Base, Ala.: Air University Press, March 2019, pp. 87–109.